A LIFE IN ERROR

The book is dedicated to the
nearest and dearest females in my life:
My wife, Rea
My daughters, Paula and Helen
My grandchildren, Stella, Anya and Elise

A Life in Error

From Little Slips to Big Disasters

JAMES REASON

Professor Emeritus, The University of Manchester, UK

CRC Press is an imprint of the
Taylor & Francis Group, an **informa** business

CRC Press
Taylor & Francis Group
6000 Broken Sound Parkway NW, Suite 300
Boca Raton, FL 33487-2742

Printed on acid-free paper
Version Date: 20160226

International Standard Book Number-13: 978-1-4724-1841-8 (Paperback)

Visit the Taylor & Francis Web site at
http://www.taylorandfrancis.com

and the CRC Press Web site at
http://www.crcpress.com

This book is like a personal and intimate trip through the ideas that pioneered human error and industrial safety. It goes into day-to-day experience of errors, contains testimonials and anecdotal information, and widens to system safety. Everything seems to have been said on the topic, and yet the book puts the matter differently in a manner that is true, full and in plain, jargon-free language. I love this book.

René Amalberti, Haute Autorité de Santé, France

Reason's new book is a master class on human error: a concise tour of his career explaining how mistakes can occur. It is a pleasure to accompany him while he presents his favourite and often funny accounts of fallibility, tempered with insights on the resulting risks and how they can be mitigated. Highly recommended as a taster text or a refresher course on error.

Rhona Flin, University of Aberdeen, UK

In this delightful memoir, Jim Reason provides an amazingly comprehensive and understandable explanation of how and why individuals and organizations make mistakes and what to do about it. A valuable review for experts and a perfect introduction for beginners.

Lucian Leape, Harvard University, USA

This book is an authoritative reminder of the journey to gain acceptance of human error as intrinsic to open systems operations as we enjoy it today, portrayed by the witty pen of one of its topmost trailblazers. I thoroughly enjoyed the book, and found the segment on organizational accidents a particular gem.

Daniel E. Maurino, formerly Coordinator of the Flight Safety and Human Factors Study Programme, International Civil Aviation Organization (ICAO)

A fascinating personal and intellectual journey showing the evolution of both James Reason's personal approach and also the broader history of thinking on error and safety. He has a unique gift for making complex ideas accessible within an absorbing and lucid narrative. And all leavened with wonderful examples of human error and some great stories.

Charles Vincent, Imperial College London, UK

Each chapter of this book tells a story where Reason personally confronted a puzzle about accidents, human performance, or organizational decisions. Together the stories build a comprehensive picture of how safety is created but sometime undermined.

David D. Woods, Ohio State University, USA

Contents

List of Figures

Foreword

It is a daunting task to write the foreword to this book—first of all, because Jim Reason probably needs no introduction to readers who have opened this book; secondly, because Jim is an undisputed master of the English language, whom many of us have as an ideal but whose level of excellence few have any realistic hope of reaching; and thirdly, because this book is an intellectual reminiscence, rather than a scientific or technical text. While it probably would be less daunting to comment on it as the latter, it would be not only boring to readers with little appetite for academic hair splitting, but also entirely inappropriate for the occasion.

Whether one likes the term 'human error' or not – and I must admit to being one of those who would like to see it wither – it is commonly used and likely to be with us for many years to come. It is therefore worth speculating a little about why that is the case. There are many practical reasons why we need the term 'error', and the book describes most of them. But in addition there is also a fundamental psychological need that was described, although probably not for the first time, about 1,000 years ago. Ibn Hazm (944–1064), considered one of the leading thinkers of the Muslim world, noted in one of his works that the chief motive for all human actions is the desire to avoid anxiety. This is a profound psychological insight. Anxiety, of course, comes in many forms, but one form follows the inability to explain or to make sense of what has happened, particularly if it was bad. This semi-pathological need of certainty creates a preference for clear and simple explanations, expressed in terms that are easy to understand and that we feel comfortable with. Error, and in particular human error, perfectly fits the bill. While this does not make it a bona fide scientific concept, it does explain why it is so popular and persistent.

This book tells an interesting story of how the current thinking about errors has developed. Although it nominally is the story of one person's intellectual development, this development mirrors, and has influenced, the understanding of safety in general and of errors in particular. The story shows the factors or events that influenced this development, from Three Mile Island (TMI) and Chernobyl to 'To Err is Human', and the progression in thinking from individual error makers to collective errors and organizational accidents, and from an information-processing orientation to one that incorporates motivational, social and institutional factors.

My own introduction to this field – and to Jim – happened in 1978 (if memory serves me well). That year I moved from Aarhus University to the Risø National Laboratory, where I started working with industrial problems (nuclear power plants, but before TMI). I quite soon stumbled on one of Jim's papers, entitled 'How did I come to do that?' (*New Behaviour*, 1975, April, 10–13). (In those days there was no Internet, no Google, and no easy access to databases, so finding an interesting paper was really a matter of stumbling.) The paper started thus: 'Absent-minded behaviour in everyday life can be comical. But the same mistakes on the flight deck of a passenger jet can end in catastrophe.' (Sadly, the paper did not contain the amusing anecdote about the cat that begins this book.) The paper was interesting and was clearly about safety issues, but I remember that I wondered whether the author's name was a pseudonym intended as a subtle joke. I soon realized that this was not the case, and shortly after had the pleasure of meeting Jim for the first time.

Jim's surname is, of course, a little ironic considering his favourite research topic, and it prevents him from achieving the eponymic fame that comes when the name of a person is used to refer to something, usually something they have produced or described. To illustrate what, just think of Pascal, Braille, Röntgen, Morse, and Richter. The word 'reason' is, of course, widely used in the English language, but it does not describe what Jim is rightly famous for, namely 'error'. Indeed, 'error' is almost the opposite of 'reason'. Reason is therefore an aptronym rather than an eponym, i.e. a name that fits a person's nature or occupation. And Jim has certainly brought reason to the study of error. So while the intellectual life that this book presents in Jim's own words may have been 'in error', it has certainly not been in vain.

Erik Hollnagel
Middelfart, Denmark, May 21, 2013

Preface

This short book covers the main way stations on my 40-year journey in pursuit of the nature and varieties of human error. Inevitably they represent a very personal perspective, but I have also sought to include contrary opinions. The journey, as at this point, begins with a bizarre, absent-minded action slip committed by me in the early 1970s—putting cat food into the teapot—and continues until the present with a variety of major accidents that have shaped my thinking about unsafe acts and latent conditions. The original focus of this enquiry was individual cognitive psychology, but over the years the scope has gradually widened to embrace social, organizational and systemic issues.

For the most part, my interest here is more on the journey than on the details of each waypoint—though there will be some exceptions. There are two reasons for this. First, many of the waypoints have been covered in previous Ashgate books. Second, I want to focus on the factors either in my head or in the world that prompted the next step in the journey.

This book is written for all those who have an interest in human factors and their interactions with the workings of technological systems whose occasional breakdowns can cause serious damage to people, assets and the environment. This is a large and diverse group whose number, I hope, includes students, academics and safety professionals of all kinds—and has lately included a growing number of health carers.

Where possible, I have tried to make clear the thinking and—if you'll excuse the unavoidable pun—the reasoning that contributed to the models, metaphors, taxonomies and practices that have influenced the course of this journey. Many of these I owe to other workers in this large field, particularly (and in no particular order) Jens Rasmussen, Don Norman, Erik Hollnagel, Rob Lee, Dan Maurino, Dave Woods, Richard Cook, David Embrey, Jan Davies, Dianne Parker, Rebecca Lawton, Najmedin Meshkati, John Wreathall, Andrew Hopkins, Carl Macrae, Patrick Hudson, Carlo Cacciabue and Jean Paries—to name but a few.

As ever, I must also express deep gratitude to my editor-in-chief, Guy Loft, for his continuing encouragement and guidance.

Chapter 1

A Bizarre Beginning

One afternoon in the early 1970s, I was boiling a kettle for tea. The teapot (those were the days when tea leaves went into the pot rather than teabags) was waiting open-topped on the kitchen surface. At that moment, the cat—a very noisy Burmese—turned up at the nearby kitchen door, howling to be fed. I have to confess I was slightly nervous of this cat and his needs tended to get priority. I opened a tin of cat food, dug in a spoon and dolloped a large spoonful of cat food into the teapot. I did not put tea leaves in the cat's bowl. It was an asymmetrical behavioural spoonerism.

I little realized at the time that this bizarre slip would change my professional life—I was a lecturer in psychology at the University of Leicester and had run out of research topics. I used to work on motion sickness and disorientation, but students did not like to be made sick and the pool of usable experimental subjects was fast drying up. I needed something new to do. I had flirted with clinical psychology, but I was neither trained nor temperamentally suited for it.

After I had washed out the teapot and sworn at the cat, I started to reflect upon my embarrassing slip. One thing was certain: there was nothing random or potentially inexplicable about these actions-not-as-planned. They had a number of interesting properties. First, both tea-making and cat-feeding were highly automatic, habitual sequences that were performed in a very familiar environment. I was almost certainly thinking about something other than tea-making or cat-feeding. But then my attention had been captured by the clamouring of the cat beyond the glass kitchen door. This occurred at the moment I was about to spoon tea into the pot, but instead I put cat food into the pot. Dolloping cat food into an object (like the cat bowl) affording containment had migrated into the tea-making sequence. Even the spooning actions were appropriate for the substance: sticky cat food requires a flick to separate it from the spoon; dry tea leaves do not. It seemed in retrospect that local, object-related control mechanisms were at work (see later).

I did not realize it at the time, but this action slip exhibited nearly all the principal characteristics of absent-minded errors:

- Both behavioural sequences were highly routine, meaning that I was unlikely to be monitoring step by step—my attention was 'absent'.
- Both the cat's bowl and the teapot afforded containment.
- Some change introduced into a routine sequence of actions (in this case, the cat's demands) quite often misdirects actions down the wrong path.

This event was of considerable psychological significance. Although unintended, the individual actions were executed smoothly and skilfully. Among other things, this error suggested a way of investigating the control of highly practiced actions. And these events occurred (in my case) on an almost daily basis and right beneath my nose. After years of laboratory study, I suddenly felt I had access to the fabric of everyday life. Experiments, by definition, are usually carried out in highly artificial (controlled) circumstances—but I felt very strongly that these naturalistic observations could reveal the stuff of which human thought and action were truly made.

I would like to end this story by describing another absent-minded slip that also took place in my Leicestershire kitchen. For once, I was not the error maker. I rarely get a chance to observe a slip in progress. I was idly watching my wife make tea. She was boiling the kettle and had the teapot beside it with the lid off. Then she reached up to a nearby shelf and took down a large jar of instant coffee. After adroitly unscrewing the lid, she put three teaspoons of coffee into the teapot, poured water into it, and was then alerted to her slip by the strong smell of coffee. Now there was nothing particularly curious in this slip. Errors of this nature happen relatively frequently in our kitchen. What made the slip especially interesting, however, was the skilful way she unscrewed the lid of the coffee jar. The tin tea caddy had a pull-off, push-on lid. The question of interest was, how did her hand perform the appropriate unscrewing movements on the lid? Her mind was clearly 'absent' from the tea-making sequence at the time. My conclusion was that her hand movements were under the local control of the coffee jar. Tea-making and most cooking are test-exit-test-exit tasks.

This raised the interesting notion that well-used familiar objects develop a local control zone. Once the hand enters that zone it is automatically directed to perform object-appropriate actions. These control zones become particularly evident when we are mooching around the house in a state of reduced intentionality—as when we are waiting for a visitor or phone call. We have no immediate goal except to pass the time. Under those conditions, we stray past a fruit bowl and pick up an apple; or we stray past the open door of a bathroom and clean our teeth, even when that has recently been done, or we front up to the toilet and take a leak that we don't really need.

These aimless periods reveal that much of our behaviour is under the control of the immediate environment, often resulting in the unintended activation of action programs appropriate for the circumstances. I have even wandered into the bedroom and started to undress in the middle of the day, standing beside my bed.

Both of these slips—the cat food and the tea-making—give us important insights into the way we control our routine actions in a familiar environment. When actions are under-specified—by reduced intentionality, forgetting, misdirected attention and the like—they tend to default to behaviours that occur frequently in that particular context. Such 'strong-habit-intrusions' occur in a variety of mental contexts.

In tip-of-the-tongue states, for example, we struggle to retrieve a word that we know we know. When this happens, we often find ourselves calling to mind a word we know, but we also know that it's not the one we seek, although it often feels 'warm'. We call these words 'blockers'. Here's an example: I was struggling find the title of the film *Deliverance*, but I kept retrieving another single-word film title:

Intolerance (the D.W. Griffiths film was much more familiar to me than *Deliverance*). Both words also had the relatively unusual '-erance' ending. Which brings us to St Augustine's question: if we keep retrieving the wrong word, how do we know it's the wrong word if we can't retrieve the sought-for word? Clearly some part of my brain knew the answer because of the '-erance' endings. This suggests that mental life is lived on several levels, and consciousness is only one smallish part of it.

Chapter 2

Plans, Actions and Consequences

Despite the fact that most people understand what is meant by the term *error*, there is no universally agreed scientific definition. Indeed, some of my distinguished colleagues would wish to dispense with the term altogether. I do not agree, so I will spend the rest of this chapter trying to put together a working definition of error that will serve as a framework for what follows.

It is generally accepted that errors entail some kind of deviation of human performance from an intended, desired or ideal standard. Although such discrepant actions may have adverse effects, they can also be inconsequential or even benign, as in trial-and-error learning or serendipitous discovery. How many of you, for example, have accidentally stumbled upon better ways of working with a computer application (e.g. a word processing package) by clicking on the wrong window and then trying out a useful command you had not used before?

Errors are not intrinsically bad, though their consequences and the workplace local conditions provoking them are often undesirable. It is not so much the psychological processes that determine the nature of the outcome; rather it is the circumstances of their occurrence that shape an error's consequences. Activating the kettle rather than the toaster in your kitchen can be mildly irritating, but the same kind of switching error committed in the control room of a nuclear power plant or on the flight deck of an aircraft can be disastrous.

Nearly all human actions, correct or otherwise, entail three basic elements: plans, actions and consequences. As such they seem a good starting point for constructing an error definition—and, as we shall see later, also an error classification.

Plans

Stop and think for a moment. What are you going to be doing for the rest of the day? Tomorrow? Next week? In six months' time? Two years hence? It is likely that you have some plans for these periods, although as the time in question extends further into the future it is probable that your answers will become increasingly more vague and uncertain. Much of what occupies our thoughts is the making and refining of plans. They are central to our understanding of error.

Consider the following imaginary example as our starting point: It is noon, and you are hungry. It's been a wretched morning and you feel like treating yourself to a decent lunch. Do you want to eat French, Indian, Italian or Chinese? You favour Italian because you have a fancy for a large plate of spaghetti carbonara. Will you go to Luigi's or La Dolce Vita? You favour Luigi's. It's further away but the parking is easier. But there is a problem: you are trying to lose weight, and spaghetti carbonara is rich in calories.

This plan begins commonly enough with the *need* to alleviate a state of tension, in this case a combination of hunger pangs and dissatisfaction with the frustrating morning. There are many ways of achieving this, but a specific *intention* is quickly formed to have a decent lunch. Again such a *goal* could be achieved by a variety of means, but one particular *plan* is favoured—to eat spaghetti at Luigi's. Having made this decision, it only remains to specify and assemble the *action sequences* that will take you to Luigi's by the most convenient route.

The plan at this stage consists of a stated aim and a rough outline of the actions necessary to achieve it. Notice that the planner does not need to fill in the fine details of each operation. These action steps are largely automatic subroutines that are implicit in the jottings already made on the mental scratchpad. They are manipulated in thinking by a series of verbal tags and mental images. The more we engage in these habitual action sequences, the fewer are the number of 'tags' required to specify them in our planning. Repetition reduces the number of low-level control statements necessary to guide our behaviour.

In the very broadest sense we can say that our actions are in error when they fail to achieve the objectives of our current plan. But here we run into a problem, as the previous example of planning indicates. Let's assume that our self-indulgent planner arrives at Luigi's without a hitch and consumes a large plate of spaghetti carbonara. In the sense that his actions fulfilled his intentions exactly, they could hardly be said to be in error. But what about his longer-term dieting plan? Viewed in this light, the successfully executed short-term plan to consume a calorie-laden dish was clearly a mistake.

The point being made here is that our lives are governed by many plans. Sometimes they nest together in close harmony, but at other times they conflict, as shown by our example. The existence of co-existing and conflicting plans would make even a working definition of error beyond our reach, were it not for two built-in limitations to human performance. First, our physical capacity to turn personal plans into action is limited. We can only be in one place at any time. In addition, we possess only a limited mental capacity for carrying out plans. Although there may be many stored plans competing for our attention, usually only one of them is maximally active at any one time.

Error, as we have seen, is not an easy notion to pin down. Dictionaries send us on a semantic circular tour through other like terms such as mistake, fault defect and back to error again. That dictionaries yield synonyms rather than definitions suggests that the notion of error is something fundamental and irreducible. But we need to probe more deeply into error's psychological meaning.

Error is intimately bound up with notions like plan, intention, action and its outcome. The success or failure of our actions can only be judged by the extent to which they achieve, or are on the way to achieving, their planned consequences.

For our present purposes, therefore, we can define error as follows:

> *The term error will be applied to all those occasions in which a planned sequence of mental or physical activities fails to achieve its desired goal without the intervention of some chance agency.*

Two qualifications are important here. Firstly, there is the inclusion of the notion of intention, and secondly, the absence of any chance intervention.

Should you inadvertently shoot a red light as the result of being stung by a wasp, you would have suffered an automatism rather than having committed an error, because the actions responsible for the undesired outcome were not what you intended or could reasonably have avoided. Unfortunately, these excuses will not wash with the police—shooting a red traffic light is a crime of absolute liability.

Similarly, if you were struck down on the street by a piece of returning space debris, you would not achieve your immediate goal, but neither would you be in error, since this unhappy intervention was outside your control. By the same token, achieving your goal through the influence solely of happy chance—as when you slice a golf ball that bounces off a tree and onto the green—could hardly be called correct performance.

The logic of this definition yields two distinct ways in which you can fail to achieve your desired objective:

- In the first case, the plan of action may be entirely appropriate, but the actions do not go as planned. These are *slips* and *lapses* (absent-mindedness) or *trips* and *fumbles* (clumsy or maladroit actions). Such failures occur at the level of execution rather than in the formulation of intentions or planning.
- The second category of failure can arise when your actions follow the plan exactly, but the plan itself is inadequate to achieve its desired goal. These are termed *mistakes* and involve more complex,

> higher-level processes such as judging, reasoning and decision-making. Mistakes, being more subtle and complex, are much harder to detect than slips, lapses, trips and fumbles. In the case of actions-not-as-planned we have a conscious record of what was intended and so the discrepancy is easily discovered. But it is not always obvious what kind of a plan would be ideal for attaining a particular objective. Thus mistakes can pass unnoticed for long periods—and even when detected they can be a matter of debate.

In the next chapter we will consider three levels of performance: skill-based, rule-based and knowledge-based. These ideas have proved extremely useful in the categorization of errors and violations and were a major step onwards.

Chapter 3

Three Performance Levels

It was argued in the previous chapter that slips and mistakes arise from quite different psychological mechanisms. Slips were said to stem from failures at the level of execution, often arising from the unintended activation of largely automatic procedural routines—also known as *action schemas*. Mistakes, on the other hand, derive from higher level mental processes involved in formulating plans, setting objectives, decision-making and judging the available information.

If that were really so, we would expect slips and mistakes to take quite different forms. But that is not the case. Both slips and certain kinds of mistakes can take very similar 'strong-but-wrong' forms, where the erroneous actions are more in keeping with past practice than the current circumstances demand. There is also another problem. Certain well-documented errors fall between the slip and mistake categories. They possess properties common to both. This is illustrated by the following errors committed by nuclear power plant operators in the United States during two separate emergencies:

- *Oyster Creek* (1979): The operators mistook the annulus level (160.8 inches) for the water level within the shroud. The two levels are usually the same. But on this occasion the shroud level was only 56 inches above the fuel elements (due to a valve closing error). Although the low water level alarm sounded three minutes into the event and continued to sound, the error was not discovered until 30 minutes later.
- *Three Mile Island* (1979): The operators did not recognize that the relief valve on the pressurizer was stuck open. The panel display indicated that the relief valve switch was selected closed. They took this to indicate that the valve was shut, even though this switch only activated the opening and shutting mechanism. They did not consider the possibility that this mechanism could have (and actually had) failed independently and that a stuck-open valve could not be revealed by the selector display on the control panel.

Neither of these errors readily fit into either the slips or mistakes category. They contain some of the elements of mistakes in that they involved incorrect appraisals of the system state; yet they also show slip-like features in that strong-but-wrong interpretations were selected. These errors can best be described as arising from an inappropriate diagnostic rule, of the kind *If (situation X prevails) then (system state Y exists).* In both cases, rules that had proved reliable in the past now yielded wrong answers in these extremely unusual emergency conditions.

And this is where Jens Rasmussen's (a very distinguished Danish control engineer) three performance levels—skill-based (SB), rule-based (RB) and knowledge-based (KB)—came to the rescue.[1] Using his framework, I was able to distinguish three distinct error types: skill-based slips, rule-based mistakes and knowledge-based mistakes.

Distinguishing the Performance Levels

Type of Activity

A key distinction between the performance levels is whether or not an individual was engaged in problem solving. Activities at the SB level involve routine and habitual action sequences with little in the way of conscious control. There is no awareness of a current problem; actions proceed mainly automatically in mostly familiar situations. But both the RB and KB levels are only triggered when the actor becomes aware of a problem—that is, when he or she has to stop and think. There are two kinds of problem: those for which you have pre-packaged solutions (RB level), and those for which you have not (KB level). Unfamiliar problems can only be dealt with by thinking 'on the hoof', usually involving trial-and-error learning.

SB slips generally precede the detection of a problem. Both the RB and KB levels are only called into play by the unanticipated occurrence of some externally or internally produced event or observation that requires a deviation from the current plan of action.

---

1 Rasmussen, J. (1983). Skills, rules, knowledge: signals, signs, and symbols, and other distinctions in human performance models. *IEEE Transactions: Systems, Man and Cybernetics. SMC* 13: 257–67.

Control Mode

Both SB slips and RB mistakes share *feed-forward control*. This emanates from stored knowledge structures (motor programs, schemas, rules). Rasmussen summarized this feature of the SB level as follows: 'Performance is based on feed-forward control and depends upon a very flexible and efficient dynamic internal world model.'[2] Comparable control processes operate at the RB level: 'Performance is goal-oriented, but structured by feed-forward control through a stored problem solving rule.' This may take the form of *If (X, Y, and Z are present) then it is (a B situation)*. Or *If (A) then (do C)*.

In contrast, control at the KB level is mainly of the feedback kind. This is necessary because the problem solver has exhausted his or her stock of stored problem-solving routines, and is forced to work 'on line', using slow, sequential, effortful, resource-limited, conscious processing. The focus of this laborious and error-provoking process will be some internalized model of the problem space. This involves setting local goals, initiating actions to achieve them and then observing the extent to which they are successful or otherwise. This process is largely error driven.

Human beings are ferocious pattern matchers. When we have exhausted our problem-solving rules, we search to find analogies. We look for a rule set that might apply to the current problem. We try very hard not to think on our feet—though, ironically, we still remain better at it than most computers. And that's why we continue to monitor highly automated systems: to deal with situations that the designers and programmers had not considered.

---

2 Ibid.

Expertise

To a large extent, human expertise resides at the RB level of performance. Medical training, for example, is largely concerned with stocking students' heads with disease- and injury-defeating rules. Experts have a much larger collection of problem-solving rules than novices, and they are formulated at a more abstract level of representation. There is a fairly close relationship between the predictability of error and the level of expertise: the more skilled an individual is in performing a task, the more likely it is that his or her errors will take 'strong-but-wrong' forms at the SB and RB levels of performance.

The Coexistence of All Three Levels

Driving is an excellent activity for demonstrating that all three levels of performance can coexist at the same time. As you drive, you can often see at the bottom edge of your visual field the movements of the steering wheel—they seem outside of your direct control. Steering and speed control operate at the skill-based level. Indeed, they can be disrupted by too much conscious attention.

The rule-based level comes into operation as soon as you drive onto a public road. Dealing with other road-users, for example, is governed by an elaborate set of rules.

Fortunately, the KB level rarely comes into play. But it can do when the radio informs you that there is a long tail back further down the road. Then you are (usually) forced to resort to what is often an inaccurate and incomplete mental map of the road system ahead in order to plan a route that will avoid the jam. It also comes into play when your mind is beset with off-road problems.

Where Next?

Chapter 4 deals with absent-minded slips and lapses. This classification has not remained fixed over the years—though I'm not sure I have improved upon the earlier versions. If any reader has either the interest or the patience to track this evolution, then let me direct you to the following books, set out in chronological order:

Reason, James and Mycielska, Klara (1982). *Absent-minded? The Psychology of Mental Lapses and Everyday Errors*. Englewood Cliffs: Prentice Hall.

Reason, James (1990). *Human Error*. Cambridge: Cambridge University Press.

Reason, James (1997). *Managing the Risks of Organizational Accidents*. Aldershot: Ashgate Publishing.

Reason, James and Hobbs, Alan (2003). *Managing Maintenance Error*. Aldershot: Ashgate Publishing.

Reason, James (2008). *The Human Contribution: Unsafe Acts, Accidents and Heroic Recoveries*. Farnham: Ashgate Publishing.

Chapter 4

Absent-Minded Slips and Lapses

Stepping into the bath with your socks on; struggling to open a friend's front door with your own latch key; squeezing shaving cream onto your toothbrush; saying 'thank you' to a stamp machine; trying to drive away without switching on the ignition; going into the wrong gender toilet; pouring a second kettle onto freshly made tea ... the list goes on, but you will have already recognized the species. We have all experienced occasions when our minds have been 'absent' from the task in hand so that our words or actions no longer run according to plan.

Similarly, we have all known times when our minds go blank and we are no longer able to retrieve from memory a word that we know to be there, lurking just beyond the reach of consciousness. Our lives are strewn with such usually inconsequential lapses. Freud called them the 'refuse of the phenomenal world'.

So if they amount to little more than psychological garbage, why should we bother with them? Especially when many believe that Freud had already taken the choicest pickings. Leaving aside the question of whether he left us anything worth finding in this mental dustbin, Freud himself provided an excellent reason for studying these apparent trivia. He wrote: 'In scientific work it is more profitable to take up whatever lies before one whenever a path towards its exploration presents itself. And then, if one carries it through thoroughly, one may find even in the course of such humble labour, a road to the study of the great problems.' By the way, my disagreement with Freud's interpretation of slips and lapses is discussed in Chapter 7, though I thoroughly endorse his comments above.[1]

The Hallmarks of the Absent-Minded (AM) Slip

An important characteristic of AM slips is that they are instantly recognizable as belonging to our personal repertoire of routine actions. We know them as our own. Clearly they are not what was intended at the time, but one can usually detect a peculiar kind of logic, especially when we stop to consider where the slip was made, what it was we intended to do and the kinds of activity we habitually carry out in those circumstances.

From an examination of many absent-minded slips, it is clear that 'act-wait-act-wait' tasks like tea-making are hugely productive of this kind of error. In my experience there's hardly a stage in the process not associated with some kind of slip. But for all that, certain types of unintended actions are never reported. We may omit to put tea in the pot, or put in a double measure, repeat the kettle-boiling sequence unnecessarily and pour tea into the sugar bowl; but is there anyone who has inadvertently tried

1 Freud, S. (1914). *The Psychopathology of Everyday Life*. London: Ernest Benn.

to blow down the spout of the teapot, or thrown a cup through the window, or tipped tea over the contents of a cupboard? These are not outside the range of behavioural possibilities; they simply don't belong in our mental store of available action programs, and don't occur as AM slips.

Nor would we use the term absent-minded to describe the faltering efforts of a beginner making her first pot of tea. She would commit many errors, but these are the kind of blunders that any novice would make. Unlike AM errors, they can take unpredictable forms. The novice's errors arise from *lack of competence*, whereas the true hallmark of the AM slip is *misapplied competence*. The beginner has not mastered the necessary actions, but her mother, with her mind on other things, has total mastery of the task. However, she has so many overlearned action sequences relating to the kitchen that the wrong one can readily be called into play, or omitted, because she is preoccupied or distracted with matters unconnected to the task.

This brings us to a key point about AM errors: they are a problem for the expert, not the novice, being characteristic of highly skilled and habitual activities. I realize that this runs counter to common sense. We acquire skills with much effort and practice in order to avoid making errors. Of course experts make fewer errors, but it's not their *quantity* that matters; it's their *type*. Although AM errors are fairly infrequent when matched against the hit-and-miss blunders of the learner, the likelihood of making such a slip increases with proficiency at a particular task. This makes them especially interesting to those wishing to learn more about how we control our largely routine actions.

Another hallmark of AM slips is that they are not random events determined exclusively by the idiosyncrasies of those who commit them, or by the place and time in which they occur. Rather they follow a clearly discernible pattern that is largely independent of the period in which the perpetrator lives. Two quotations will support the claim that AM errors are a timeless and universal human characteristic.

The first comes from the seventeenth-century French essayist Jean de La Bruyère, who described the antics of one his contemporaries, the Comte de Brancas:

> He plays at backgammon and asks for something to drink; it is his turn to play, and having the dice box in one hand a glass in the other, being very thirsty, he gulps down the dice and almost the box as well, throwing the liquor on the board and half drowning his antagonist.

A second quotation is a charming account of an AM slip contributed to the *Spectator* in 1711 by an English journalist, Mr Eustace Budgell. He described how he was walking in Somerset Garden with his friend, Mr Will Honeycomb. During the course of the walk, Mr Honeycomb picked up a small pebble whose shape appealed to him. Mr Budgell again:

> After we had walked sometime, I made a full stop with my face towards the west, which Will knowing it to be my usual method of asking what's a' clock in an afternoon, immediately pulled out his watch and told me we had seven minutes good before our club time. We took a turn or two more, when, to my great surprise, I saw him fling away his watch a considerable way into the Thames, and with great sedateness in his looks put up the pebble he had before found in his fob.

Although the details are different, both slips were of the same kind: behavioural spoonerisms—just like my cat food fiasco. The difference here is, of course, that both the slips described in the quotations were symmetrical spoonerisms when the two objects get reversed—as opposed to my slip, which was one-sided only. Errors like these, involving complete or partial reversals, are very commonplace today. Familiar examples are unwrapping a sweet, putting the paper in your mouth and discarding the sweet; or striking a match, putting the spent match in your pocket and throwing away the matchbox.

There is a close resemblance between action slips and the errors found in speech. The behavioural spoonerisms committed by the Comte de Brancas and Will Honeycomb are clearly of the same type as those attributed to the Reverend W.A. Spooner, who is purported to have said such things as 'queer old Dean' when he meant to say 'dear old Queen'. Sometimes the outcome can be very embarrassing, as when one spoonerizes the phrase 'plucking pheasants'.

Although embarrassing and bizarre, AM slips are the penalty we pay for having a human mind. To be more specific, they are the price we pay for being able to devolve the control of our habitual actions to lower-order automatic routines. Life would be insupportable if we were constantly *present-minded*, having to make separate conscious decisions about every small act. If that were so, we would never get dressed in the morning, having spent so long in deciding how we should tie our shoes.

What Goes Absent in Absent-Mindedness?

Conscious concerns, whether internally or externally generated, consume the major part of the limited attentional resource during waking hours. In addition to continual moment-to-moment variations in the total amount of this resource that is available, the quantity drained off by these conscious concerns differs according to their nature and intensity. All mental and physical activities, no matter how automatic they may appear, make some demands on attention. The more habitual the activity and the more invariant the environment in which it occurs, the smaller is this demand. But it is always present in some degree.

One feature of absent-minded errors, as the term itself suggests, is that most slips and lapses occur when a large part of this resource has been 'captured' by something other than the task in hand. If our highly routine actions depart from intention because the limited resource is being employed elsewhere, the obvious conclusion is that, on those particular occasions, a greater degree of attentional involvement is necessary to ensure the desired outcome. I am not suggesting that a fixed amount is required throughout. But there are occasions, particularly at choice points where a familiar sequence branches into a variety of well-trodden paths, when a larger attentional investment is necessary.

Since schemas (knowledge structures in long-term memory) appear capable of being activated independently of current intentions—by needs, emotions, context, associations with other schemata and the frequency and recency of past use—some part of the attentional resource is always being consumed to restrain those activated schemas not required for our current plans.[2] The more activated these unwanted schemata are, the more of the attentional resource will be consumed in suppressing them. The execution of any activity requires the correct sequencing of several necessary schemas, and since each of these transitions between schemas might potentially lead in many directions, some attentional supervision is needed to keep them on the right track. To do this it is necessary not only to select, but also to inhibit those pre-programmed action sequences that seek to usurp this position.

If schemata were merely passive entities, like computer programs, which act only on orders from above, this problem would not arise. But schemas behave in an energetic and highly competitive fashion to try to grab a piece of the action. Our evidence suggests that when the attention resource is largely claimed by something other than the immediate activity, it is this suppressive function that is most likely to fail. In short, this appears to be what goes absent in absent-mindedness.

Strong Habit Intrusions

Over 40 per cent of all AM slips are of this kind. They take the form of intact, well-organized sequences that recognizably belong to some activity other than the one that is currently intended. This other activity is judged as being recently and frequently performed, and as sharing similar locations, movements and objects with the intended actions.

---

2 Reason, J. (1990). *Human Error*. New York: Cambridge University Press.

In addition to these general disposing conditions, there are at least four further situations in which strong habit intrusions are likely to occur:

- *When a change of goal demands a departure from some well-established action sequence.* For example, you decide to lose weight and so wish to eliminate putting sugar on your cereal. But when you come to sit down for breakfast, you automatically sprinkle sugar on your cereal.
- *When changed local conditions require the modification of some oft-performed action sequence.* As an experiment, my wife and I decided to reverse the positions of two adjacent drawers in our kitchen. One of them held our cutlery, the other didn't. It took more than four months before we stopped trying to get our knives and forks out of the wrong drawer.
- *When a familiar environment associated with particular routines is entered in a state of reduced intentionality.* For example, we could be waiting for a phone call or a visitor, then we could stray into the bathroom and clean our teeth, though this was neither necessary nor intended.
- *When features of the present environment contain elements that are similar to those in highly familiar circumstances.* For example, as I approached the turnstile on my way out of the library, I pulled out my wallet as if to pay—although I knew no money was required. I behaved as if I was at the checkout counter in a supermarket.

In the next chapter, we review questionnaire studies that sought to measure individual proneness to absent-mindedness.

Chapter 5
Individual Differences

This chapter deals with the use of self-report questionnaires to obtain individual estimates of individual proneness to AM slips and lapses. Since the mid-1970s a variety of these questionnaires have been used by different research groups. This work has been extensively reported elsewhere.[1] Rather than cover old ground, I will summarize some eight years of research carried out at the University of Manchester with the Short Inventory of Minor Lapses (SIML).

Can self-report questionnaires produce robust and consistent findings? Have they told us anything new? And do they have any practical utility? I will try to show that the answers to all three questions are 'yes', though sometimes in unexpected and unusual ways.

The Short Inventory of Mental Lapses (SIML)

The SIML is a 15-item questionnaire comprising general descriptions of the most frequently occurring slips and lapses. The instructions require the respondent to rate how often each type of cognitive

1 Reason, J. (1993). Self-report questionnaires in cognitive psychology: Have they delivered the goods? In A. Baddeley and L.Weiskrantz, *Attention: Selection, Awareness and Control*. Oxford: Oxford Science Publications.

failure has occurred over the previous six months. Five response categories are offered for each item: hardly ever, sometimes, quite often, frequently and nearly all the time. These are scored on a 1 to 5 scale and summed to give a total score. The individual items are listed below in the order in which they appear in the SIML.

1. How often do you forget to say something you were going to mention?
2. How often do you have the feeling that you should be doing something, either now or later, but you can't remember what it was?
3. How often do you find your mind continuing to dwell upon something that you would prefer not to think about?
4. How often do you find you can't remember what you have just done or where you have just been (e.g. when walking or driving)?
5. How often do you leave some necessary step out of a task (e.g. forgetting to put tea in the teapot)?
6. How often do you find that you can't immediately recall the name of a familiar person, place or object?
7. How often do you think you're paying attention to something when you're actually not (e.g. when reading a book or watching TV)?
8. How often do you have the 'what-am-I-here-for' feeling when you find you have forgotten what you came to do?
9. How often do you find yourself repeating something you've already done or carrying out some unnecessary action (e.g. switching on the light when you're leaving room in daylight)?
10. How often do you find you've forgotten to do something you intended to do?

11. How often do you decide to do something and then find yourself side-tracked into doing something different?
12. How often do you find yourself searching for something that you've just put down or are still carrying around with you?
13. How often do you forget to do something that you were going to do after dealing with an unexpected interruption?
14. How often do you find your mind wandering when you're doing something that needs your concentration?
15. How often do you make errors in which you do the right actions but in relation to the wrong objects (e.g. unwrapping a sweet, throwing the sweet away and putting the paper in your mouth)?

The SIML was tested on two independent samples. Our aim was to obtain norms and to see how stable the answers to these questions were. The two samples were as follows:

- Sample A was made up of 543 people. Our intention was to make the group as variegated as possible. They comprised 112 male and female undergraduates (mean age 20 years, ranging from 18–45); 88 female patients undergoing treatment for breast cancer (mean age 52 years, ranging from 24–85); 225 elderly females (mean age 65, ranging from 50–90); 118 elderly males (mean age 67, ranging from 50–93).
- Sample B was made up of 1,656 car drivers, comprising 847 males and 809 females. The age range was from 17–91 years.

There was a close similarity between the means and the standard deviations for these two independent samples. The correlation between them over the 15 items was 0.879. This concordance clearly satisfies the criterion of robustness. Whatever the SIML is measuring, it is evidently doing so consistently across the 15-item profile. In both cases, the most commonly occurring slips were failing to recall a name, forgetting to say something, forgetting intentions and mind-wandering. The least common were transposition errors, omitting a planned step and repetitions. There were no gender differences, but there were some large and unexpected age differences.

Age Differences

Something that has cropped up stubbornly in all of our SIML studies is the counterintuitive finding that the self-reported incidence of minor slips and lapses diminishes with age. These age differences were highly significant in both Sample A and B. There were, however, three significant reversals of this trend in both samples. These occurred on Item 6—difficulty in recalling names. The older the subject, the more frequently he or she experienced difficulty in remembering people's names. This result is consistent with William James's observation that 'when memory begins to decay, proper names are what go first ....' With the exception of Item 5—omitting necessary task steps—and Item 12—searching for something carried or just put down—all the other items show a highly significant decline with age ($p < .0001$).

In another study, using the comparable Cognitive Failures Questionnaire (CFQ) devised by Donald

Broadbent's Oxford group, it was found that 50-year-olds reported significantly more slips and lapses than did either 60-year-olds or 70-year-olds. The latter two groups did not differ significantly. In a second study, Rabbit and Abson (1991) found significant differences between all three age groups. The mean CFQ score for 50-year-olds was 40.6; for 60-year-olds it was 37.4; and for 70-year-olds it was 34.5.[2]

There are two ways of interpreting these puzzling data: either one can accept that these self-report ratings truly represent the behavioural reality, or one can challenge the validity of the questionnaire instruments, as some have done.[3] But neither is an easy path to take for the reasons set out below.

If one accepts that these self-report frequency ratings actually correspond with the respondents' everyday cognitive performance, then there are at least two possible explanations:

- The *activity* hypothesis: This argues that older people are less active and therefore they encounter fewer opportunities for making AM errors.
- The *compensation* hypothesis: One version of this is that older people, being aware of their diminished cognitive competence, rely more heavily on memory aids and reminders and hence suffer fewer lapses. Another version that we can call the 'dread' variant is that whereas younger people take AM slips and lapses as a fact of life, older people

2 Broadbent, D. et al. (1982). The Cognitive Failures Questionnaire (CFQ) and its correlates. *British Journal of Clinical Psychology*, 21: 1–16.

3 Rabbit, P. and Abson, V. (1990). 'Lost and found': some logical and methodological limitations of self-report questionnaires as tools to study cognitive ageing. *British Journal of Psychology*, 81: 1–16.

> may view them as harbingers of Alzheimer's disease and feel much more anxious about them; as a consequence they invest a greater degree of attention in the performance of everyday routines and so make fewer AM errors.

But neither of these explanations accords with the large differences between the young (mean age 22 years, range 17–29) and the middle-aged (mean age 39 years, range 30–49) found in Sample B. In the first place, it is highly unlikely that there will be major differences in activity levels between these younger groups. Second, aging studies have found no significant loss of cognitive competence before the age of 50. In addition, on the one occasion in which we sought to test the 'dread' variant of the compensation hypothesis, we found no significant differences between older and younger people in the degree of anxiety associated with each of the 15 SIML items.

Rabbitt and Abson[4] challenged the validity of self-report questionnaires on a number of counts. They argued, for example, that reliance on total scores masked the domain specificity of everyday memory skills. But this cannot apply to the SIML, since the judgements were made in regard to very specific kinds of AM lapses.

They also suggested that the lower scores of older people arise from the fact they are likely 'to forget that they forget'. This may be so, but the force of the criticism would be more compelling if the respondents were asked to gauge the actual numbers of AM lapses. They actually judged their frequency on a 5-point ordinal scale.

Another possibility relates to the fact that there may be systematic age differences in the way people perceive

4 Ibid.

the response scale. Viewed from the perspective of a long life, categories such as 'quite often' and 'frequently' may have a different significance than they do for younger people. There are two reasons to doubt this. First, respondents were asked to judge the frequency of their cognitive failures over the past six months rather than over an entire lifetime. Second, the SIML-item difference between the young and the middle-aged were greater than those between the middle-aged and the elderly.

At present, therefore, none of these criticisms are particularly convincing. The only solid thing is the age-related decline in reported slips and lapses. This remains a difficult thing to explain, or to explain away. It is likely that this uncomfortable result will continue to bother age researchers for some time to come.

The Stress-Vulnerability Hypothesis

An especially important finding from self-report questionnaires is the relationship between a high level of cognitive failures and the number and degree of psychiatric symptoms experienced during or immediately following a period of real-life stress. Donald Broadbent has argued that high rates of AM slips and lapses are related to increased vulnerability to real-life stressors. Researchers have now confirmed these findings over a number of stresses: examination stress; the stresses associated with homesickness among university undergraduates; and stresses associated with breast surgery. I will discuss the latter further below.

Among life's nastier stress-provoking knocks, there can only be a few worse than discovering a breast lump requiring surgery. In a three-year study at the University Hospital of South Manchester,

100 women were interviewed and tested around the period of their surgery. Sixty-seven of these patients had breast cancer and 33 had benign breast lesions. Sixty of the ladies with cancer and 29 with benign disorders were interviewed and tested again four months after their surgery. Both sets of interviews included the administration of a variety of paper-and-pencil measures, including the SIML and Goldberg's General Health Questionnaire, designed to measure the presence of minor psychiatric symptoms.[5]

At the first interview, the SIML correlated significantly with the number of psychiatric symptoms reported at the time. In addition, depressed patients had higher SIML scores. In addition to supporting the findings of Broadbent's Oxford group, these findings further highlight the relationship between proneness to AM slips and lapses and depression.

Analysis of the findings at the four-month testing period focused on possible predictors of psychiatric morbidity. The dependent measure was the number of psychiatric symptoms yielded by Wing's Present State Questionnaire.

Nine factors accounted for over 35 per cent of variance in a multiple regression analysis. The following ranking of predictors was obtained by the method of elimination (the figures in parentheses indicate the relative contribution of each factor):

- Upset by scar (24.2)
- SIML score (18.6)
- Radiotherapy (14.9)

5 Reason, J. (1989). Stress and cognitive failure. In *Handbook of Life Stress, Cognition and Health* (ed. S. Fisher and J. Reason), pp. 405–21. Chichester: Wiley.

- Pain in scar (8.0)
- No confiding partner (7.9)
- Previous psychiatric problems (1.6)
- Children under 14 (0.7)
- Mastectomy (0.5)
- Difficulty in moving arm (0.3)

Thus we had a remarkable finding: a marked degree of absent-mindedness in the year preceding the surgery was a better predictor of psychiatric morbidity than a number of more obviously relevant medical and social factors. The test-retest reliability of the SIML score between the first and second testing sessions was 0.69.

These findings (and other stress-related testing) provided strong support for the stress-vulnerability hypothesis. But what are the underlying mechanisms? One tenable view is that people differ characteristically in the way they deploy their limited attentional resources in the face of competing demands. This appears to influence the selection of coping strategies to deal with the stress. Error-prone individuals seem to select more resource-intensive strategies and so are less able adjust them to the discretionary aspects of the situation than are less error-prone people.

In conclusion, it is not so much that stress induces a high rate of cognitive failure (though that may indeed be so); rather it is that a certain style of cognitive resource management can lead to both absent-mindedness and to the inappropriate matching of coping strategies to stressful situations.

So far, we have been dealing with individual differences in AM slips and lapses in a very statistical (nomothetic) fashion. In the next chapter, we will take a much more person-oriented (idiographic) approach when we come to examine a forensic application of the SIML to offer a legal defence against the criminal charge of shop theft.

Chapter 6

A Courtroom Application of the SIML

In the early 1980s, I was retained by a solicitor to act as an expert witness in the defence of a man charged with shop theft—let's call him Mr X—who vehemently denied the intent to steal and said that he had absent-mindedly overlooked the unpaid-for items.[1]

He was charged with leaving his local supermarket on two occasions, separated by a week, with goods that had not been paid for. Although he was charged with the offences, he was not stopped on the first occasion, but items seen in the front section of the trolley did not appear on the till receipt. On the second occasion, when he was stopped by the security staff, the disputed items were again in the front section of the trolley. He apologized and offered to pay for the items, but the offer was not accepted.

It did not help that the offence was committed twice, though on both occasions he had bought and checked through a large number of items held in the rear part of the trolley. Were his actions commensurate with AM behaviour? I believed they were on three counts: first, it was a familiar setting; second, it was a hot day and he had taken and partially drunk a can of fruit juice

1 Reason, J. (1993). Self-report questionnaires.

and, on arrival at the checkout, he was preoccupied with ensuring that the partially empty can was paid for. Third, the disputed items were in a separate part of the trolley. These three factors—familiarity, preoccupation or distraction, and separation—are all hallmarks of AM behaviour.

This view was strongly supported by Mrs X, who stated that her husband was an exceedingly absent-minded man and wrote a detailed account of his recent slips and lapses.

I interviewed Mr X some months after these events. My main purpose was to assess his level of absent-mindedness using the SIML. The SIML was completed twice, once by Mr X in regard to himself and once by Mrs X, who was asked to rate her husband's proneness to each of 15 items. Both tests were carried out in isolation, with each unaware of the other's responses, either then or later. Neither was forewarned of the testing, though both readily agreed to it. The procedure was modelled on Donald Broadbent's practice of using 'close others' to check upon the validity of a person's ratings.

Mr X's SIML score provided strong support for his wife's view that he was a very absent-minded man. His total score was in the 95th percentile for Samples A and B. Mrs X's score put him slightly higher than this. There was also a very close correspondence in their score profiles over the 15 items. And both profiles revealed a very idiosyncratic pattern quite unlike the mean profiles for the normative samples.

This unusual pattern of Mr X's responses and its close correspondence with Mrs X's ratings, a correlation of 0.64, showed that both sets of scores related to the same person; and this, in turn, strongly suggested that both respondents were being truthful. But would this be enough to convince a jury, given the subjective nature of these ratings and the fact that a high AM score would be to Mr X's obvious advantage in pleading his case? It was almost certain, therefore, that the prosecuting counsel would direct part of his cross-examination to casting doubt on this conclusion.

One way of resolving this problem was to discover what people (of comparable age and intelligence to Mr X) did when they were directly instructed to fake their responses. Twenty-six management students completed the SIML after being told to imagine that they had been accused of shop theft and were, in fact, guilty, but had offered absent-mindedness as a defence. They were then asked to inflate their judgements, knowing it was to their advantage to obtain a high score. The results are shown in Figure 6.1.

The first thing to note is that the fakers' mean profile corresponds very closely to the normative one (correlation +.77), but is elevated over the whole range of items. This suggests that the management students approached their task by first asking themselves how they would normally respond to each item and then adding a fairly constant 'faking factor'.

From Figure 6.2, it is clear that Mr X's idiosyncratic profile did not fit the faking pattern of responses. The correlation between the profiles was only 0.11. On items 9, 14 and 15, Mr X's scores were approaching one standard deviation below those of the normative sample.

Clearly, no one of these pieces of evidence was sufficient to establish Mr X's innocence; but they added up (together with the circumstantial factors) to

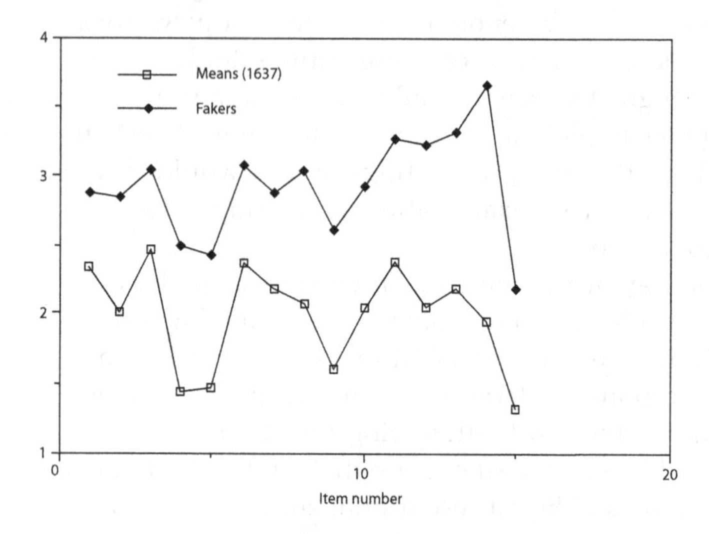

Figure 1 Comparing the mean SIML item scores for the fakers (n=26) with the sample B norms (n=1637)

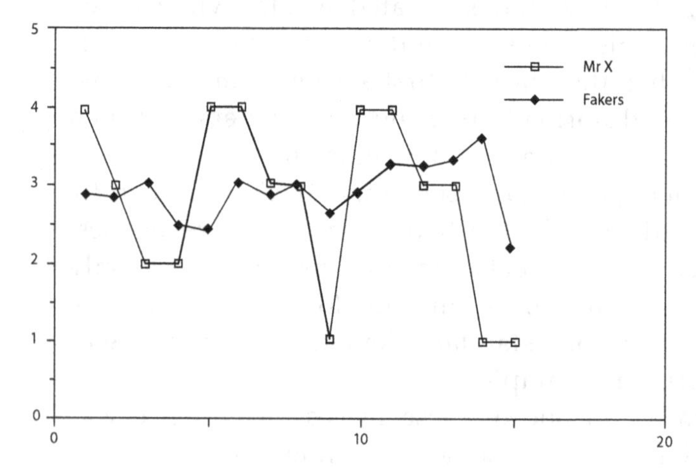

Figure 2 Comparing Mr X's SIML profile with that for the fakers

a 'reasonable doubt' that his actions might have been absent-minded rather than felonious.

Mr X emerged as a man acutely aware of his proneness to distraction and absent-mindedness, but who had worked hard to improve his powers of concentration for professional purposes. His almost cat-like ability to sustain a narrow focus of attention was crucial for his work, but very nearly proved to be his undoing in the supermarket.

Postscript: AM in Shops

Some years later, we analyzed letters written by 67 people (courtesy of the Portia Trust) who believed that they had been wrongly accused of shop theft. Fifty-three per cent of this group specifically cited absent-mindedness, mental 'blanks', or confusion as the primary cause of their predicaments. Sixty-nine per cent mentioned being distracted or preoccupied when the incident occurred. Twenty-three per cent were receiving medical treatment at the time, and 50 per cent were involved in negative life events—divorce, separation, discovery of a spouse's infidelity, bereavement, ill children and the like. For nearly a third of the accused, these life crises were present in truly diabolical combinations.

This suggested that most of these critical lapses occurred while the shopper's limited attentional resource was heavily engaged with something other than the immediate task. In a significant proportion of the letter-writing sample, this attentional fixedness was likely to have been exacerbated by medication.

A further contributing factor, mentioned by many of the accused, was the use of some unwise practice.

This mostly involved transferring goods from the store's trolley or basket to a personal receptacle. This was done to separate items of different sizes, or for different people, to make carrying more convenient or simply through inadvertence. The shopper then failed to declare the goods at the checkout.

At the same time, we carried out a questionnaire study of absent-mindedness in shops.[2] A total of 150 men and women were asked how often they had experienced each of 24 varieties of lapses while shopping. Six of these errors (risky lapses) could lead to charges of shop theft. The remaining 18 items were relatively trivial in their consequences. The most important finding was that the six risky items were judged as having occurred far less often than the non-risky items. This suggested that, under more normal circumstances, shoppers—being well aware that they could be observed by security staff—deliberately avoid actions that could be interpreted as being furtive or suspect. Thus, they appear to allocate some part of the attentional resource to guard against the making of potentially risky errors.

Nonetheless, 18 per cent of the sample judged that they had, at some time, exited a shop without paying for some goods.

Approximately half of the sample also completed the Cognitive Failures Questionnaire (CFQ). A positive and significant correlation was found between the CFQ and the Absent-Mindedness in Shops Questionnaire (AMSQ). A factor analysis of the AMSQ results revealed a very pervasive general

---

2 Reason, J. and Lucas, D. (1984). Absent-mindedness in shops: its correlates and consequences. *British Journal of Clinical Psychology,* 23: 121–31.

factor, together with a risk appreciation factor. These results supported the view that responses to both the CFQ and AMSQ reflected individual differences in the management of some high-level attentional control resource.

The moral is clear: shops can be very dangerous places. Expect no mercy from the retailer. It was also the case that people who had been wrongly accused of shop theft subsequently developed even more severe psychiatric symptoms. There are times when not even your nearest and dearest will buy the AM excuse—the retailer almost certainly won't. Be on your guard!

Chapter 7

The Freudian Slip Revisited

We should begin by reminding ourselves of what Freud actually wrote about slips and lapses. He was rarely one to mince his words: 'A suppression of a previous intention to say something is the indispensable condition for the occurrence of a slip of the tongue.'[1]

A slip is a product of both a local opportunity and a struggle between two mental forces: some underlying need or wish and the desire to keep it hidden. Freud applied similar arguments to slips of action and memory lapses. Indeed, it was his inability to recall the name of a minor poet that set him on the track in the first place. And therein lies his genius: his ability to see the value of what he called 'the refuse of the phenomenal world'.

Freud was well aware of alternative explanations. He called them psycho-physiological factors, a label that embraced fatigue, excitement, strong associations, distraction, preoccupation and the like. He was even willing to concede in a half-hearted way that a few slips could occur for these reasons alone: 'we do not

1 Freud, S. (1922). *Introductory Lectures on Psycho-Analysis.* London: George Allen & Unwin.

maintain that every single mistake has a meaning, although I think that is very probable'.[2] To Freud, notions such as absent-mindedness, excitement or distraction offered little or nothing in the way of real explanation.

> They are mere phrases ... they facilitate a slip by pointing out a path for it to take. But if there is a path before me does it necessarily follow that I must go along it? I also require a motive determining my choice, some force to propel me forward.[3]

A Classic Example

One whole chapter of *The Psychopathology of Everyday Life*[4] was devoted to a single slip. Freud regarded this analysis as one of the most convincing demonstrations of his thesis. It also reveals him as the travelling companion from hell.

On a holiday trip, Freud met a young man who was bemoaning the troubles of Jews in the Austro-Hungarian Empire. The young man quoted—or attempted to quote—a line from Virgil in which the spurned Dido seeks vengeance on Aeneas. What the young man actually said was: '*Exoriare ex nostris ossibus ultor*' ('Let an avenger arise from my bones'). He was immediately aware that he had got the quote wrong and foolishly asked the all-too-willing Freud to explain why it had happened.

Freud began by giving the correct quotation: '*Exoriare aliquis nostris ex ossibus ultor*' ('Let someone arise as an avenger from my bones'). He then asked

---

2 Ibid., p. 22.

3 Ibid., p. 36.

4 Freud, S. (1914). *The Psychopathology of Everyday Life*. London: Ernest Benn.

the unfortunate young man to free-associate on the missing word, *aliquis* (someone). His responses went as follows: dividing the word into *a* and *liquis*; relics, liquefying; fluid; saints' relics; Saint Simon, Saint Benedict, Saint Augustine and Saint Januarius (the latter two were calendar saints). Saint Januarius's miracle of blood (a phial of his blood is supposed to liquefy once a year). Finally, he got to the crunch—the fact that he was very worried that his girlfriend back in Vienna had missed her last period.

One can imagine the smug expression on the maestro's face at this moment. The important clues, according to Freud, were the allusions to calendar saints and the idea that blood flows on a certain day. Even the choice of quotation had a meaning. Dido was crying out for her descendants to avenge her—a clue to the young man's equally fervent desire that no descendants should be in the offing.

Should we give Freud the expected round of applause? Probably not, since the textual critic, Sebastiano Timpanaro, devoted a whole book to providing an alternative explanation, albeit a far more mundane one.[5]

Timpanaro pointed out that the boy's misquotation contained two separate errors: it omitted the pronoun *aliquis*, and the words *nostris* and *ex* had been reversed. He then argued very convincingly that, to a young man of his classical education, both the presence of the word *aliquis* in the sentence (which is redundant anyway—it means 'someone') and the correct order of *nostris* and *ex* are highly unusual forms (nowhere else does Virgil use this particular ordering).

---

5 Timpanaro, S. (1976). *The Freudian slip*. London: NLB.

The sentence is thus susceptible to the process of banalization: the replacement of archaic or unusual expressions with forms that are in more common use.

Slips Undone

Let's take another of Freud's examples and see if the same rather unexciting counterarguments could also apply. His friend Dr William Stekel told him of an embarrassing incident that had occurred when he was saying goodbye to a female patient after a house call. Stekel extended his hand to the lady and then discovered to his horror that it was undoing the bow that held together her loosely fastened dressing gown. Stekel commented: 'I was conscious of no dishonourable intent, yet I executed this awkward movement with the agility of a juggler.'[6]

To Freud, of course, the interpretation was obvious. Stekel harboured unprofessional desires for the woman, a secret betrayed by his unwitting hand movements. But there is also a more boring explanation. Stekel, momentarily distracted or preoccupied, had fallen into a habit-plus-affordance trap. Nineteenth-century medicine was a very hands-on affair. In the course of his house calls he would have been accustomed to undoing the bows of bed jackets and the like to palpate a patient's chest or abdomen. A strong habit was thus established and bows naturally afford untying. All that was required to trigger the gaffe was some wayward attention just prior to the intended hand-shaking sequence.

---

6 Freud, S. *The Psychopathology of Everyday Life*, pp. 136–7.

Duller Alternatives

With these two examples, we have assembled most of the ingredients for an alternative view. There are, in my view, at least two necessary conditions for provoking an absent-minded error. Firstly, some cognitive under-specification—inattention, incomplete sense data or insufficient knowledge; secondly, the existence of some locally appropriate response pattern that is strongly primed by its prior usage, recent activation or emotional charge and by the situational calling conditions. There is also the prediction that an error sequence is likely to be more familiar, more frequent and more typical in context than the intended correct sequence. We can test out these ideas on an instance of what, on the face of it, was a classical Freudian slip.

Some years ago, I attended the opening of a new building designed to house clinical psychologists. The person who made the opening speech was a local politician. After extolling the virtues of clinical psychology at some length, she concluded by saying: 'I declare this Department of Cynical—er, I mean Clinical Psychology open.'

At first I thought, as Freud probably would have done, that that was what she really meant. But there were duller alternatives. First, 'cynical' and 'clinical' have very similar structures and sounds and would therefore fit equally well into the articulatory programme. Second, the word 'cynical' anticipates the initial phonology of 'psychology'—the first syllables have the same sound. How often have we heard newsreaders make similar anticipatory errors? Third, being a politician, it is likely that that the word 'cynical' had far more currency in her everyday lexicon than did 'clinical'.

The claim here is not that Freudian slips do not occur. They almost certainly do. The argument is about their relative frequency. Modern psychologists[7] would contend that most everyday slips and lapses have more banal origins—along the lines indicated above. But there is a simple test that can be applied. For a slip to be convincingly Freudian, it should take a less familiar form than the intended word or action.

A cursory search of Freud's own examples yielded one strong possibility—that of the Viennese lady who, when calling her children in from the garden, said 'Juden' (Jews) instead of 'Jungen' (boys). It is likely that for any mother, the latter word would have been used more frequently than the former, even if she were Jewish herself.

Conclusion

So, was Freud right about Freudian slips? Rightness, particularly in psychology, is not an all-or-nothing thing. He was probably wrong in asserting that all (or nearly all) slips are, in some way, intended. But if we ask whether Freud was correct in his view that slips represent minor eruptions of unconscious processing, then the answer would be an emphatic 'yes'. But we would not necessarily take the strict psychoanalytic interpretation of 'unconscious'; rather it would be one that relates to processes that are not directly accessible to consciousness. The large part of mental life—automatic processing—falls into this category.

7 Norman, D. (1981). Categorization of action slips. *Psychological Review*, 88: 1–15.

Slips, we now believe, provide important glimpses into the minutiae of skilled or habitual performance. But they can also reveal suppressed feelings. Freud was perfectly correct in refusing to divorce cognition from emotion.

On balance, therefore, he was mostly right—though he would hate the 'mostly'.[8]

8 I am deeply indebted to the editor of *The Psychologist*, who permitted me to use large parts of an article that had appeared in 2000, vol. 13, 12.

Chapter 8

Planning Failures

What follows is just a joke, but it always makes me laugh. In addition, it captures many of the defining features of a mistake, particularly the inadequacy of situational awareness. Here goes:

> A struggling and mostly out-of-work actor auditioned for a part, and got it. The director explained what it comprised: 'As soon as the curtains open, you will say your single line—Hark, hark, the cannons roar. Got it?' The actor said he had. No problem! He was told the play would not open for a couple of months and that, given the limited nature of the role, he was not required for rehearsals. For the subsequent eight weeks, he went about his daily business muttering the line to himself over and over again. Hark, hark, the cannons roar. Hark, hark, the cannons roar. On opening night, he stood poised behind the closed curtains. When they opened he was faced with the audience and the theatre was filled with a huge banging and booming. He fell to his knees, put his hands over his head, and howled 'Oh my God! Whassat?'

In Chapter 2, the basic characteristics of mistakes and how they differed from actions-not-as-planned were outlined. In Chapter 3, I spelled out the arguments leading to the distinction between rule-based and knowledge-based mistakes. This chapter focuses on the ways in which planning can go wrong—at both the individual and collective levels of its formulation.

Planning

The elements that come together in the mental activity of planning do not fall readily into Rasmussen's skilled-based (SB), rule-based (RB) and knowledge-based (KB) typology. Where they fit depends largely on their complexity and the degree of uncertainty that exists about the future. The future is always uncertain, but the extent to which it may be is also a function of the nature of the activity and the associated environment. Where this is relatively trivial or commonplace, as in our earlier example (see Chapter 2) of choosing what and where to eat lunch, the basic elements could be spat into consciousness by long-term knowledge schemas, and the actions required to achieve the goal could be derived from well-established action schemas. Both of these might be highly routinized and hence operate at the SB level.

But in this chapter, I am interested in more complex plans that may work at the RB level, but more likely, they involve a mixture of the RB and KB levels, or even—given big surprises—solely at the KB level.

A Brief Sketch of the Planning Process

It is suggested that the processes involved in planning involve three major components: a limited attentional resource (a working database); a set of mental operations that act upon this database; and the schemas (or schemata), or the knowledge structures stored in long-term memory. Also assumed, but not specified in any detail, are the input (receptors) and output functions (effectors) linking the planner to the world.

The Working Database

This is limited in capacity and continuously variable in content. At least three kinds of information are likely to be contained within it:

- That derived directly from the environment via the input function.
- That which has been called up from the schema base.
- That which has been spontaneously thrown up by active, though not necessarily plan-relevant, schemas.

The Mental Operations

Three inter-linked operations are involved: selection, judgement and decision-making. The contents of the working database may be selected by the planner both from the immediate environment and from the long-term schema store. The judgements are of two kinds: those related to goal setting and those concerned with goal achievement. Finally, decisions are made as to the goal and the actions by which it is to be attained. Thus, the planning process may be broken down into four stages:

- Setting the objectives.
- Searching for alternative courses of action.
- Comparing and evaluating alternatives.
- Deciding upon the course of action.

The Schemas

Schemas (or schemata) are involved in all stages. They contribute both selected and uncalled-for (spontaneously 'spat' up) information to the working database. The latter will comprise conscious fragments (images and words) thrown up by highly activated schemas. These are likely to include emotionally charged material, stored information triggered by the local situation or through association with plan elements, or outputs from recently and frequently used schemas. It is likely that the completed plan is stored as a set of sequentially linked schemas that are primed to go into action at some appropriate time or when the right circumstances are met.

Possible Sources of Bias Leading to Planning Failures

These are grouped under three headings, each corresponding to one of the components of planning.

Sources of Bias in the Working Database

- At any one time, the working database will 'display' only a small fraction of the information potentially available and relevant to the current plan.
- Of the several variables that might bear upon the planning process, not more than two or three of them are likely to be represented in the database at any one time.
- The planning process is unlikely to be sustained for more than a few seconds. And when it is resumed, the contents of the database will be different.
- Planning is more shaped by past experiences than is appropriate, given the possible variability of future events.
- The information that is 'called' into the database will be biased in favour of those items emanating from activated schemas, something that may be more potent than the relevance of the information to the plan.

- The information present in the working database will be biased in favour of past successes rather than past failures.

Sources of Bias in the Mental Operations

- Planners being guided primarily by past events will underestimate the influence of chance. As a result they will plan for fewer contingencies than are likely to occur.
- Planners will give more inferential weight to information according to its vividness or emotional impact.
- Planners are heavily influenced by their theories: where these are inappropriate, systematic sources of bias will be introduced into the planning process. One consequence is that planners will unconsciously fill in the bits missing from the current evidence in accord with their theories.
- Planners are not good at assessing population parameters on the basis of data samples. They have little appreciation of the unreliability of small samples or of the effects of bias in the sampling procedures.
- Planners are poor at detecting many types of covariation. They are disposed to detect it only when their theories predict it.
- Planners will be subject to the 'halo effect'. They show a predilection for single orderings and an aversion to discrepant orderings. They have problems processing two separate orderings of the same people or objects. As a result they reduce these discrepancies to a single ordering by perceived merit.

- Planners tend to have a simplistic view of causality. They tend to believe that a given event can only have one sufficient cause. They also suffer from 'hindsight bias'—knowledge of the outcome biases their assessments of the prior events.
- Planners tend to be over-confident in evaluating the correctness of their knowledge.

Schematic Sources of Bias

- These are most likely to show themselves after the planning has been completed and before the plan is executed. One very potent source of bias stems from a strong urge to seek confirmatory evidence for the soundness of the plan and to disregard information that suggests the plan may fail.
- A completed plan is not only a set of directions for future action; it is also a theory about the future state of the world. It confers order and reduces tension. It will be strongly resistant to change. This unwillingness to change is likely to be greater when the plan is complex, has been the result of much time and effort, and entailed the involvement of many people.
- Dominating these features is a strong urge to make sense of all the plan features—it is what has been called 'effort after meaning'. It is so strong that we are prepared to be mistaken in certain instances so long as we can continue to preserve order within the world at large.

Collective Planning and Its Failures

Most of the plans that impinge upon our lives are the product of organizations and groups.

Organizational Planning

An influential theory of collective planning was developed by Herbert Simon. It was called 'Behavioural Theory of the Firm' and founded in large part upon Simon's principle of bounded rationality:

> The capacity of the human mind for formulating and solving complex problems is very small compared with the problem whose solution is required for objectively rational behaviour in the real world....

This limitation gives rise to 'satisficing' — the tendency to select satisfactory rather than optimal courses of action. Organizational planners are inclined to compromise in their goal-setting by choosing minimal objectives rather than those yielding the best possible outcomes.

Richard M. Cyert and James G. March questioned the omniscient rationality that was assumed to exist on the part of the planners. Rather, they argued that the planning process was best understood in terms of general rules of thumb — known as heuristics. Four such heuristics were identified:

- Quasi-resolution of conflict
- Avoidance of uncertainty
- Problemistic search
- Selective organizational learning

Downs elaborated a theory of decision-making within large organizations. He identified four self-serving biases common to all officials:

- Each official tends to distort the information he passes upward in the hierarchy, exaggerating those data favourable to himself and minimizing those that are unfavourable.

- Each official is biased in favour of those policies or actions that advance his own interests and programmes.
- Each official will vary the degree to which he complies with directives from above, depending on whether these directives favour or oppose his own interests.
- The degree to which each official will seek out additional responsibilities and accept risks will vary directly with the extent to which the initiative will help him achieve his personal goals.

Other biases depend upon the nature of the official. 'Climbers' are strongly motivated to invent new functions for their departments, and to avoid economies. 'Conservers', on the other hand, are biased against any changes in the status quo. Officials that have been in a post a long time tend to become 'conservers'. The middle levels of the hierarchy are likely to contain more 'conservers' than either the lowest or highest levels. The more a bureau relies on formal rules, the more conservers it is likely to contain.

Although organizations are subject to more sources of planning failure than individuals, there are some similarities in the underlying error tendencies:

- Organizations, like individuals, plan on the basis of very limited databases.
- Other biases, common to both organizations and individuals, can be traced to the over-use of labour-saving heuristics.

- Both show themselves to be prisoners of past experience in their preference for well-established routines rather than novel departures.

This tendency of both organizations and individuals to err in a conservative direction was neatly expressed by Barbara Tuchman in her comments on the Schlieffen Plan: 'Dead battles, like dead generals, hold the military mind in their dead grip, and Germans, no less than other people, prepare for the last war' (p. 38).[1]

Planning in Small Groups: 'Groupthink'

The American social psychologist Irving Janis studied the decision behaviour of small, cohesive and often elite planning groups. His reasons for examining such groups were that '... all the well-known errors stemming from the limitations of an individual and of a large organization can be greatly augmented by group processes that produce shared miscalculation'. He called this 'groupthink'. They were mostly high-level groups involved in US foreign policy. They ranged from the Cuban Bay of Pigs fiasco to the escalation of the Vietnam War.

The 'groupthink' syndrome was characterized by eight main symptoms:

- An illusion of invulnerability, creating extreme optimism and the willingness to take excessive risks.
- Collective efforts to rationalize away warnings that might have led to reconsideration of the plan.

1 Tuchman, B. (1962). *The Guns of August*. London: Four Square Books.

- An unswerving belief in the rightness of the group's intentions.
- Stereotyped perception of the opposition as being either too evil to negotiate with or too stupid to counter the planned actions.
- The exertion of group pressure on any member that deviated from the collective stereotypes, illusions or commitments.
- Self-censorship of any doubts felt by individual members.
- A shared illusion of unanimity, arising both from self-censorship of doubts and the assumption that silence means consent.
- The emergence of self-appointed 'mindguards'—members who saw it as their duty to protect the group from any contrary opinions or adverse information.

These group dynamics appear to add an emotional dimension to the distortion of the planning process, and this serves to bring individual biases into greater prominence but also adds some specific biases to group membership. The powerful forces of perceived togetherness act in concert to render the possibility of failure unthinkable—and if not unthinkable, then certainly unspeakable.

This chapter offered a somewhat different 'take' on mistakes than those discussed elsewhere in the Ashgate list. In Chapter 9, we move from errors to violations—quite a different phenomenon, but they are coupled together with errors under the general heading of 'unsafe acts'. But they are not always unsafe; sometimes they lead to salvation.

Chapter 9

Violations

The Chernobyl disaster occurring over the Easter weekend of 1986 was a landmark event by anyone's standards, but—aside from living among radioactive sheep—it had a very special significance for me. A close examination of the operators' behaviour leading to the explosion of the reactor's core revealed two quite distinct types of unsafe act. There was an unintended slip at the outset of the fatal experiment that caused the reactor to operate at too low a power setting where it was subject to xenon poisoning. Unable to bring the power level up to the desired 20 per cent, the operators deliberately persisted with the trial and in so doing made a serious violation of safe operating procedures. They did this partly because they did not really understand the physics of the reactor, but also because of their determination to continue with the testing of the voltage generator—which, ironically, was intended as a safety device in the event of a power loss.

This was an important way station because it marked the beginning of a major widening of the scope of my enquiries. Up to this time the focus had been very largely upon individual error makers. But the appearance of violations required a shift away from a purely cognitive information-processing orientation to one that incorporated motivational, social and institutional factors, and paramount among the latter were rules and procedures.

We do not generally mean to make slips, lapses or mistakes. Errors arise from informational problems either in the mind of the individual or in the world at large. Violations, on the other hand, are deliberate acts—that is, people usually mean to break the rules or fail to comply with procedures, though they generally do not intend the bad outcomes that these deviations sometimes bring about (unless they are terrorists). Violations arise largely from motivational factors, from beliefs, attitudes and norms and from the organizational culture at large. Culture was a very important contributor to the Chernobyl catastrophe. Each Soviet reactor operated in isolation. There was little or no sharing of safety information. Accidents often happen because people forget to be afraid. But the Chernobyl operators did not forget, rather they had never learned to be afraid.

I have discussed violations at length in other Ashgate publications, so I will only cover the topic in outline here. I will also be stressing occasions when violations have had happy rather than unsafe outcomes. By the same token, there are some instances when strict compliance has led to disaster.

Violation Types

We have found it convenient to distinguish four types of violation:

- *Corner-cutting or routine violations*: These are committed to avoid unnecessary effort or to circumvent clumsy or inappropriate procedures. Many organizations have two sets of procedures—

the official ones and those that get written down in 'black books'. These informal procedures derive from skilled and experienced operators who recognize that the official procedures are often written by those who have never carried out the task in question. There is often a conflict between those who design a task and those who perform it; gradually, the two usually come together and a workable set of directions evolves. Here is an example of a corner-cutting violation derived from aircraft maintenance:

- A 747 aircraft was about to make its first flight after a major service in which the oil lines on one engine had been changed.
- Finding oil leaks on an engine run, the engineers checked and tightened the suspect oil lines.
- However, the engineers skipped an additional engine run because the tug had arrived.
- A subsequent oil leak from the engine caused an in-flight shutdown and diversion.

- *Thrill-seeking or optimizing violations*: We have many, often conflicting, needs. Nowhere is that more apparent than when we are driving a motor vehicle. Our primary aim is to get from A to B, but en route, we sometime seek to satisfy some baser urges—the joy of speed, cutting in, tailgating and the like. In this respect, there are marked demographic differences that are not apparent in error-making: males violate more than females, the young violate more than the old. To a large extent, it depends on whether the driver is playing to an audience of his passengers. Women, by and large, are not impressed by violating; hence they can act as a restraint on driver violations.

- *Necessary violations*: Many organizations continue to write additional procedures—usually to proscribe the human actions that were implicated in the last accident or event. The result is that the scope of allowable action is less than that necessary to get the job done. We have noticed this particularly in work upon the railway infrastructure. I stated earlier that if there is one general principle for the production of error it is *under-specification* of the processes necessary for correct action. But in necessary violations the problem is the direct opposite: it is *over-specification*. Here the responsibility lies with the system rather than with the individual worker. The danger is that if this is not rectified, the violations necessary to get the job done become routinized and form a regular part of the workers' repertoire of actions.
- *Exceptional violations*: As the term suggests, these one-off events are likely to occur when the plant or vehicle is operating under exceptional conditions. It has been noted that nuclear power plant events are most likely to occur during startup or shutdown. These are a normal part of the plant's operation, but they occur infrequently relative to steady state running and so operators are less practiced and the procedures often sparse. Sometimes they have to 'wing it'. Such a case occurred at Chernobyl. Among other things, the test procedures required a dangerous violation: the switching off of the emergency core cooling system, and the repeated shutting off of defences, became a defining part of the sad history of the accident. The next example is exceptional in quite a different way. Two engineers were tasked with inspecting an oil pipe. One of

> them jumped down into an inspection pit and was rendered unconscious by noxious fumes. His partner, driven by the need to give him speedy aid, also jumped down and was similarly overcome. Dangerous inspection pits were well known and the golden rule was, in case of an accident, radio for help, don't jump in. But this didn't take account of the human imperative to help a friend in trouble.

The 'Mental Economics' of Violating

The German psychologist Petra Klumb[1] investigated the costs and benefits of non-compliance. Where the latter outweigh the former, a violation is almost certain to occur. The perceived 'balance sheet' is shown below:

> *Perceived benefits*: An easier way of working; saves time; more exciting; gets the job done; shows skill; meets a deadline; looks macho.

> *Perceived costs:* Possible accident; injury to self or others; damage to assets; costly to repair; risk of sanctions; loss of job; disapproval of friends.

The benefits of non-compliance are immediate and the costs are remote from experience: violating often seems an easier way of working and for the most part brings no bad consequences. In short, the benefits of non-compliance are often seen to outweigh the costs.

The task here is not so much to increase the costs of non-compliance (by tougher sanctions and stiffer

1 Klumb, P. (1988). The economics of violating. Personal communication.

penalties), but to try to increase the perceived benefits of compliance. That means having procedures that are for the most part workable, correct and available. They should describe the quickest and most efficient ways of doing the job. Any lack of trust caused by inappropriate or clumsy procedures will increase the perceived benefits of violating. And sometimes, as we have already seen, the task can only be done by bending the rules.

I mentioned earlier that violations do not always have bad outcomes. In many cases, significant innovations and improved practices can only be achieved by those prepared to push the boundaries, and in some cases to exceed them.

Breaking the Rules and Winning

The widely used expression 'turning a blind eye' had its origins in the events that occurred on 2 April 1801. On that day, Vice Admiral Horatio Nelson was leading a fleet of 12 British warships into Copenhagen harbour. Operating under the command of Admiral Sir Hyde Parker, Nelson was charged with bringing the Danes to heel as part of destroying the League of Armed Neutrality. This alliance of Russia, Prussia, Denmark and Sweden cut the British off from desperately needed naval supplies around the Baltic.

Unable to see the action clearly, Admiral Parker thought that Nelson was getting the worst of the fight and signalled (by flags) for him to withdraw. Nelson placed the telescope to his blind eye and said to his flag captain, Sir Thomas Foley, 'You know, Foley, I only have one eye—I have the right to be blind sometimes. I really do not see the signal'. The battle continued, with Nelson destroying the bulk of the Danish fleet before a truce was negotiated. Following the Battle of Copenhagen, the League of Armed Neutrality was disbanded.

This was not the last time that Nelson bent the rules to his advantage. At the Battle of Trafalgar in October 1805, the combined French and Spanish fleets had emerged from Cadiz harbour with the intent of chasing off the British fleet that had been bottling them up. The convention of naval warfare at that time was that opposing fleets should line up, side by side, and blast away with their broadsides. But Nelson had other ideas. He put his fleet in a line astern configuration and sailed directly at the enemy lines at a right angle to them with the intention of splitting them. It was a hard fought battle that Nelson eventually won, leaving the Royal Navy master of the seas for more than a century to come.

Of course, this brings us to an obvious conclusion: In order to break the rules or flaunt convention and still triumph, the violator has to be very good at what he does. Had Nelson lost either or both of these battles, it would have been quite a different story.

Nelson was an aggressive and talented risk taker, as was General Robert E. Lee, commander of the Confederate Army in the American Civil War. Lee won many battles, but it was the Battle of Chancellorsville in the spring of 1862 that stands out most in my mind. Lee was confronted by a Federal Army that was superior to his in manpower, weaponry and infrastructure (the Union had almost a monopoly on industrial resources). One of the basic rules of war is: don't divide your troops in the face of a larger force. But Lee sent Stonewall Jackson and his division on a 16-mile hike to the far end of the Federal Army line while he engaged them at the near end. It was a bloody battle, but Jackson and Lee managed to roll up the northern army and send it scurrying back home.

With such an imbalance in resources, Lee had to rely on his very considerable wits and southern fighting skills in order for the Confederacy to survive. But even the best of generals has his bad days—as at Gettysburg in the following year.

Here is one more war story: David and Goliath. The Philistine army was lined up along one side of a ravine with the Israelites facing them on the other side. Neither moved for 40 days. But a very large and fearsome Philistine, Goliath—reputed to be over eight feet tall—walked up and down in front of his battle line throwing jeers and taunts and challenges at King Solomon's army. David, his teenage son, volunteered to fight Goliath. The king was reluctant to let him go, but was eventually persuaded. He offered David his sword and armour. David began to dress himself for battle, but then stopped. He realized that if he came within sword-swiping distance of Goliath, he was dead meat. So he filled his pouch with round stones and advanced on Goliath whirling his sling. He loosed a stone which struck the giant on the forehead and killed him. Convention dictated that you fought a single combat fully armed. David chose not to. He didn't and he won.

A Bittersweet Finale

Not that there was anything sweet about Piper Alpha, the focus of this concluding section, except that there were some survivors. It was a sad case of demonic perversity. Those—the majority—who died followed the procedures; those—the few—who lived did not, mostly out of necessity.

On 6 July 1988, an oil and gas platform, Piper Alpha, operated by Occidental Petroleum, exploded.[2] The explosion and the resulting oil and gas fires killed 167 men; there were only 61 survivors. Most of the deaths occurred on the top platform. The safety regulations required all personnel to assemble by the accommodation in the event of an emergency. Unhappily, this position put them in the line of a plume of smoke and flames. The death toll included two members of a rescue vessel.

After the accident, I spoke to one of the divers. He told me that he and his colleagues, already low down on the structure, jumped into the sea and were picked up by rescue boats.

The total insured loss was £1.7 billion. At the time of the disaster, the platform accounted for approximately 10 per cent of North Sea oil and gas production. It was the worst offshore disaster in terms of lives lost and industry impact. The subsequent inquiry by Lord Cullen made recommendations that radically changed the emergency procedures and their regulation in a wide variety of oil-related industries.

2 The Hon. Lord Cullen (1990). Public inquiry into the Piper Alpha Disaster. London: Department of Energy, HMSO.

Chapter 10

Organizational Accidents

The dozen years between 1976 and 1988 were marked by a succession of grisly major disasters worldwide, most of them manmade (see list below). They were also the years that I was developing the notions of 'organizational accidents' and latent failures—later to be modified to latent conditions.

Major Accidents 1976–88 (a very abbreviated list)

- 1976 – Seveso
- 1978 – Tenerife runway crash
- 1979 – Three Mile Island
- 1981 – Hyatt walkway collapse, Kansas City
- 1982 – Sinking of Ocean Ranger submersible drilling unit
- 1983 – Colombo, Sri Lanka
- 1984 – Bhopal
- 1986 – Chernobyl
- 1986 – Challenger spacecraft
- 1987 – King's Cross underground fire
- 1988 – Piper Alpha oil and gas rig explosion

Organizational Accidents

Although these disasters differ widely in location, technology and costs, they all share at least three characteristics:

- Many of the contributing factors were present within the system before the actual catastrophe occurred, sometimes for many years. This observation gave rise to the notion of latent failures. These 'resident pathogens' are present in all high-tech systems.
- All of the systems listed above had multiple defences, barriers and safeguards designed to prevent known hazards from coming into damaging contact with people or assets.
- The disasters occurred because an unforeseen concatenation of latent conditions—human unsafe acts and local triggers—defeated the many defences, creating a trajectory of accident opportunity, causing damage and loss.

One representation of these views that has been widely used in aviation and nuclear power generation is the 'Swiss cheese' (SC) model—my claim to 'fame'. It is discussed at length elsewhere,[1] so I will only give a brief sketch here.

In the Swiss cheese metaphor, the system defences are represented as slices of cheese. They intervene between the operational hazards and potential losses and victims. In an ideal world, each defensive layer would be intact, but, in reality, they are more like Emmenthaler cheese, having many holes. But unlike

1 Reason, J. (2008). *The Human Contribution*. Farnham: Ashgate.

the holes in Swiss cheese, these holes are in constant flux, opening and shutting and shifting their location. The presence of holes in any one 'slice' does not normally bring about a bad outcome. This can only happen when the holes in successive layers line up to permit a pathway of accident opportunity. Where a system has 'defences-in-depth', these accident opportunities occur very rarely, but their consequences can be extremely damaging.

The holes in the defences can arise for two reasons: *active failures* and *latent conditions*. Nearly all organizational accidents involve a complex interaction between these two sets of factors.

Active failures are the errors and violations committed by those in direct contact with the system. Their impact upon the defences is generally short-lived: a brief opening and shutting of a hole. Other holes are longer lasting. System designers, builders, procedure writers, maintainers and system managers create such latent conditions unwittingly because none of these people can foresee all the possible scenarios of system failure. Latent conditions may lie dormant for many years before they combine with active failures and local triggers to create an accident trajectory.

Unlike active failures, whose specific forms are hard to predict, latent conditions can—in theory—be identified before a bad event occurs. This proactive form of safety management involves regular monitoring of the system's 'vital signs'—generic systemic processes such as planning, scheduling, training, communicating, designing, building, operating and maintaining. A number of techniques created for this purpose have been described in detail elsewhere. The aim of such measures is to enhance the intrinsic 'safety health' or resilience of complex technological systems. The purpose is to make systems more resistant to their operational hazards. One of the key features shaping a system's robustness is its culture, to be discussed in the next chapter.

When an organizational accident occurs, the key question is not who blundered, but how and why did the defences fail? Quite often, but not always, such enquiries reveal proximal unsafe acts. Should they be identified, the next question is 'what workplace factors contributed to these unsafe acts?' This is likely to identify things such as work pressure, inadequate training and/or briefing, under-staffing, inappropriate tools and equipment, unworkable procedures, time of day, poor supervision and the like.

These local provocative conditions are likely to be the product of decisions made by top-level management. Such decisions may prove to be mistaken, but that is not always the case. Almost all high-level decisions are likely to carry a penalty for someone, at some time, somewhere in the system. For instance, one of the tasks of senior managers is to allocate resources among the departments of the organization. Such a distribution is rarely made on a purely equitable basis. There are often good reasons why one department should receive a larger slice of the resource 'cake'. But it is not always appreciated that the smaller allocations can translate into error-provoking conditions in the workplace.

The Error Contribution

It has been estimated that human error is implicated in some 70 to 80 per cent of all accidents involving hazardous technologies. The human factor thus appears to be the weakest link. But it was not always so. In the 1960s the estimated contribution of human error was around 20 per cent. Have human beings become more fallible in the past 40 years? Not necessarily. There are a number of reasons why the error contribution is now more salient than hitherto.

- The material and mechanical elements of hazardous technologies have become markedly more reliable. This has reduced the proportional contribution of technical failures, thus making the 'human error' category more evident.
- The years from the mid-1970s to the late 1980s brought about a widespread increase in the power and diversity of computers. As a result, the level of automation increased dramatically. One consequence has been that more and more supervisory control has been exercised by fewer and fewer operators (centralization).
- More automation can have three consequences. First, it can place an increased cognitive burden on the individual, particularly at times of high work load. Second, layers of computing make the system more opaque to those who control it—particularly in the first generation of Airbus aircraft when the system had been pre-programmed to disallow certain manoeuvres, often without the pilots' knowledge. Third, although automated control

reduces the occurrence of slips and lapses, it actually places a greater load upon the operator's reasoning skills when something goes wrong. While automation may actually reduce the total number of errors, it increases the probability that, when errors do occur, they will be mistakes. These are both harder to detect and likely to do greater damage to the system because they affect safety-critical issues such as problem-solving and decision-making. Automation does not eliminate error; it merely relocates it, often to a more dangerous sphere of operations.

Contrasting Two Kinds of Accidents

It has been argued at length elsewhere[2] that accidents fall into two groups: individual and organizational accidents. They have contrasting properties as indicated below:

- *Individual accidents*: These are frequent and usually have limited consequences (lost time, injuries or damage to assets). They occur in systems where there are relatively few defences. They arise from limited causes: slips, trips and lapses. They have a relatively short history.
- *Organizational accidents*: Rare occurrences that have widespread and often devastating consequences. They occur in systems that have many layers of defences, barriers, safeguards and controls. Their contributing factors arise at many levels of the system and often extend back in time—

2 Reason, J. (1997). *Managing the Risks of Organizational Accidents*. Farnham: Ashgate.

sometimes for many years. These 'big bangs' are low frequency/high severity events: explosions, crashes, collisions, collapses and releases of toxic substances. A key question is: do individual accidents provide a reliable guide to a system's vulnerability to organizational accidents? The answer is an emphatic 'no'. The road to disaster is frequently paved with declining or low lost time injury frequency rates (LTIFRs). Here are some examples:

- Westray mining disaster (1992), Canada. Twenty-six miners died. The company had just received an award for reducing the LTIFR.
- Moura mining disaster (1994), Queensland. Eleven men died. The company had halved its LTIFR in the four years preceding the accident.
- Longford gas plant explosion (1998), Victoria. Two died, eight injured. Safety efforts directed at reducing LTIFR. Major hazards of unrepaired equipment not recognized.

British Petroleum's US Refineries Independent Safety Review Panel stated the problem very succinctly in its discussion of the causes of the Texas City explosions in 2005:

> BP primarily used injury rates to measure process safety performance at its US refineries before the Texas City accident. Although BP was not alone in this practice, BP's reliance on injury rates significantly hindered its perception of process risk.

Individual and organizational accidents have different causal sets. Organizational accidents arise from a concatenation of linked failures of multiple system defences. Individual accidents, on the other hand, involve the failure of personal protection against injury, where these can be mental or physical. Common to both kinds of accidents are commercial pressures, inadequate resources and an inadequate safety culture. Culture is the topic of the next chapter, and especially the organizational factors that resist cultural change.

Chapter 11
Organizational Culture: Resisting Change

As we saw in the previous chapter, the 1980s seem to have had more than their 'fair share' of organizational accidents. One of the many consequences of this was that the 1990s became the *safety culture decade*. Though there was not—nor is—any agreed definition of culture, there was a strong belief that tightly coupled, complex, well-defended computer-based technologies—nuclear power, chemical process, transport systems and the like—are somehow more susceptible to the adverse effects of a poor safety culture than traditional industries involving close encounters between people and hazards—as in mining, construction, the railway infrastructure and road transport.

Because of their diversity and redundancies, the defences-in-depth will be widely distributed throughout the system. As such, they are only collectively vulnerable to something that is equally widespread. The most likely candidate is safety culture. It can affect all elements in a system for good or ill.

Since the mid-1990s, there have been a large number of well-attended meetings and workshops devoted exclusively to organizational safety culture. A notable example was the US National Transportation Safety Board (NTSB) symposium on 'Corporate Culture and Transportation Safety'. This meeting attracted some 550 delegates from all of the NTSB's constituencies: aviation railroads, highways, the marine world, pipelines and the safety of hazardous materials. The symposium was convened because the NTSB's accident investigators were increasingly conscious of the crucial role played by cultural factors in contributing to bad events.[1]

There are at least three ways in which an inadequate safety culture can adversely undermine a complex system's protective layers—the slices in the Swiss Cheese Model (SCM).

1. A poor safety culture will increase the number of defensive weaknesses, or 'holes', created by active failures—the errors and violations of those at the 'sharp end'. Such unsafe acts are more likely to occur in organizations that neglect to identify and remove recurrent error traps. But, perhaps more dangerously, a poor safety culture will encourage an atmosphere of non-compliance with safe operating procedures. Such violations not only increase the likelihood of a subsequent error being made, they also make it more probable that such errors will have damaging consequences.
2. An inability to recognize and respect the full extent of the operational hazards can lead to the creation of more and longer-lasting holes in the defensive layers. These can arise as latent conditions during maintenance, testing and calibration, or through

1 Reason, J. (1998). Achieving a safe culture: Theory and practice. *Work & Stress*, 12: 293–306.

the provision of inadequate equipment, or by downgrading the importance of training and briefings in dealing with emergencies. All of these factors have featured in recent organizational accidents.

3. Perhaps the most insidious and far-reaching effects of a bad safety culture are shown by the organization's reluctance to deal proactively with known deficiencies in the defences, barriers and safeguards. In short, defensive gaps will be worked around and allowed to persist. Investigations of organizational accidents provide abundant examples of management neglecting or postponing the correction of previously identified defensive weaknesses.

The Barings Bank Example

Some months before the collapse of Barings, the London-based senior management noticed that tens of millions of pounds remained unaccounted for in the dealings of their Singapore futures trading office. (The 'rogue trader', Nick Leeson, had previously arranged for the computer-based reporting system to be 'doctored' so as to conceal his massive losses.) The Group Treasurer was later to acknowledge that there was no excuse for not making the balancing of these books the highest priority. 'But', he said ruefully, 'there was always something else more pressing to do'. This statement could stand as an epitaph for a large number of culture-induced organizational accidents. If there is a phrase that captures the essence of an unsafe culture, it is *unwarranted insouciance.*

What Makes a Safe Culture?

According to Karl Weick and his colleagues, the power of a safe culture lies in instilling a 'collective mindfulness' of the many entities that can penetrate, disable or bypass a system's safeguards. This is no easy task in industries with relatively few accidents.[2]

In a telling phrase, Weick described reliability (and safety) as a 'dynamic non-event'.[3] Safety (as opposed to 'unsafety') is invisible in the sense that uneventful outcomes do not attract attention. If nothing bad happened yesterday, then people will often assume that nothing bad will happen today if they go on acting as before. But this is dangerously misleading because it fails to take account of the many unrecorded adaptations, adjustments and tweaks performed by front-line operators to keep (what is almost always) an imperfect system operating safely. Human beings can be heroes as well as hazards, a fact that is largely ignored in a research methodology that is triggered primarily by the occurrence of adverse events.

I have written at length on organizational safety culture in a number of Ashgate books, so I will not pursue the definitional issues further. Instead, I want to focus on attempts to change culture and particularly upon the individual and organizational factors that act against these changes.

---

2 Weick, K.E., Sutcliffe, K.M. and Obstfeld, D. (1999). Organising for high reliability processes of collective mindfulness. *Research Into Organizational Behavior*, 21: 23–81.

3 Weick, K.E. (1991). Organizational culture as a source of high reliability. *California Management Review*, 29: 112–27.

Gradations of Cultural Change

There are many steps along the road to attaining a safer culture, and many opportunities for failure. At any one time an organization may be in any of the states set out below. This list represents a continuum of progress along the path of change. Of the eight change conditions shown below, only the last involves a successful and enduring transition. You might care to ask yourself whereabouts does your organization lie on this continuum.

1. *Don't accept the need for change*; the system managers are happy with the status quo. They do not believe they have a safety problem, and they are satisfied with the ways they are achieving their cost-saving and efficiency targets.
2. *Accept the need for change, but don't know where to go*. There is concern over a series of bad events. There is unwelcome media comment. Existing safety measures are recognized as inadequate, but the cultural deficiencies are not understood or appreciated.
3. *Know where to go, but don't know how to get there*. Acknowledge that the existing safeguards are less than adequate. Recognize that the organizational culture is not conducive to safer operations, but unsure how to make the necessary improvements.
4. *Know how to get there, but doubt whether the organization can afford it*. Current projects are overrunning their budgets. There is a high turnover of skilled labour and shortages of key staff.
5. *Make changes, but do them only cosmetically*. Take short cuts. Fail to validate the process. Fail to monitor progress.

6. *Make changes, but no good comes of them.* The model for the changed organization does not align with the real world.
7. *Model aligns today but not tomorrow.* The change achieves only limited benefits due to unforeseen or unappreciated changes in the external world.
8. *Successful transition.* The changed organization keeps in step with a dynamic world and brings sustained benefit.

Vulnerable System Syndrome

It usually takes a fair measure of bad luck for the 'holes' in a well-defended system to line up to permit an accident opportunity. But for all that, there still remain a number of characteristics that make a system especially prone to having an organizational accident. Accident investigations in various domains suggest that a cluster of organizational pathologies—the vulnerable system syndrome (VSS)—render some systems more liable to adverse events.

VSS has three interacting and self-perpetuating elements: blaming front-line operators, denying the existence of systemic error-provoking features, and the blinkered pursuit of the wrong kind of excellence—usually productive and financial indicators. The imperative need to achieve these targets is often cited as the reason why necessary systemic safety improvements can't be made.

This vulnerable system syndrome is present in some degree in all organizations. Seeking for and recognizing its presence and taking effective remedial action are essential prerequisites of cultural improvement and effective safety management.

Blame and denial interact and potentiate one another. Together they are the most pernicious and dangerous of the core elements of VSS. Collectively, they conspire to ensure that those whose business it is to manage system safety will generally have their eyes firmly fixed on the wrong ball. There are strong drivers at both the personal and the organizational levels. These are reviewed below.

The Dynamics of Blame and Denial

Both blame and denial have strong psychological 'drivers' that operate at both the personal level and the organizational level. Factors functioning at both levels are summarized below.

At the Personal Level

- *Fundamental attribution error*: This is one of the main reasons why some people are so ready to accept human error as an explanation rather than as something that needs explaining. When we see or hear of someone performing less than adequately, we tend to put it down to his or her personality or ability. We say that the person was careless, stupid, thoughtless, irresponsible, incompetent or reckless. (This is especially so in health care, as we shall see in the next chapter.) But if you were to ask the actor why they did what they did, they would almost certainly point to the circumstances and situational constraints. Everyone is capable of a wide range of behaviours, sometimes ill-judged, sometimes inspired, but mostly somewhere in between.

- *The illusion of free will*: People, especially in western cultures, place great value in the belief that they are largely the controllers of their own destinies. Feeling ourselves to be capable of choice naturally leads us to assume that other people are the same.
- *The 'just world' hypothesis*: This the belief shared by most children and some adults that bad things only happen to bad people and vice-versa. It is also the case that people are judged by the severity of the outcome.
- *Hindsight bias*: This is the universal tendency to view past events as somehow more foreseeable to the people on the spot than they actually were. When we look back at some salient event, our knowledge of what happened unconsciously colours our perceptions of how and why it occurred. Those blessed with outcome knowledge see all the lines of causality homing in on some clearly defined happening, but those equipped only with foresight do not necessarily see this convergence. This is one of the reasons why what seem to us to be obvious warning signs are often ignored. But such warnings are only effective if the participants know what kind of bad event they are headed for, and this is rarely the case.
- *Outcome bias*: This is the tendency to evaluate prior decisions according to whether the outcome was good or bad. A common belief is that bad events can only arise as the result of bad decisions, and conversely. But history teaches us otherwise: good decisions can still have bad outcomes.

At the Organizational Level

- *Shooting or discounting the messenger*: The American social scientist Ron Westrum distinguished three kinds of safety culture: pathological, bureaucratic and generative.[4] The main distinguishing feature is the way in which an organization handles safety-related information. Pathological organizations are liable to shoot the messenger and ignore or deny the information. Bureaucratic organizations (the large majority) listen to the message but do not necessarily know what it signifies. New ideas often present problems; safety management tends to be compartmentalized. Failures are isolated rather than generalized. Generative (high-reliability) organizations welcome the messenger, even rewarding him or her, and treat the message very seriously. They go beyond the local circumstances to make remedial changes to the system as a whole. Generative organizations assume that bad events will occur; they are driven by chronic unease.
- *Principle of least effort*: It is often fairly easy to identify the proximal unsafe acts of the person at the sharp end, and to consider these to be the 'cause' of the mishap. That being the case, investigation of the adverse event need proceed no further. And this has a corollary: the belief that anyone who says differently is a disloyal troublemaker. Blaming thus fosters denial.
- *Principle of administrative convenience*: By restricting the search to the actions of those directly in contact

4 Westrum, R. (1996). Safety of a technological system. *NTSB Symposium on Corporate Culture and Transportation Safety*. 24–5 April, Crystal City, VA.

with the system, it is possible to limit the blame accordingly. They accept that there may be the occasional 'bad apple', but they believe the barrel (the system) is in good shape. Having 'dealt with' the wrongdoer, it is a very short step to believing that it could not happen here again.

- *Entrapment*: Another process that sustains denial is what has been called a 'culture of entrapment'.[5] This analysis was based on a public inquiry into children's heart surgery at the Bristol Royal Infirmary (BRI). Between 1988 and 1994, the mortality rate at this hospital for open-heart surgery was approximately double the rate of any other centre in England for most of this period. The inquiry concluded that neither the senior clinicians nor their managers acknowledged that the hospital's performance in paediatric surgery was inadequate. Why? The inquiry report described BRI as a collection of 'tribes', fragmented, loosely coupled, self-contained subcultures. Within each subculture there is a high degree of autonomy. But this autonomy worked against learning. The poor results of the paediatric surgeons were both conspicuous and irrevocable. They were rationalized: 'It's a new field of surgery'; 'We're on a learning curve'; 'Our cases are unusually complex'. Weick and his colleagues summarized it very elegantly: 'Through repeated cycles of justification, people enact a sensible world that matches their beliefs, a world that is not clearly in need of change'. What begins as a plausible rationalization hardens into dogma because it

5 Weick, K.E. and Sutcliffe, K.M. (2003). Hospitals as cultures of entrapment. *California Management Review*, 45: 73–84.

makes sense of the world and reduces personal vulnerability. For these reasons, however, it also precludes learning and improvement.

- *Organizational silence:*[6] Many people have found themselves in situations where they have felt unable to speak up or pass issues of concern upwards. The dynamics that create and maintain silence will be hard to change, partly because a loss of employee trust is not easily reversed. Creating a climate that encourages employees to speak up requires some fairly dramatic changes: a new pluralistic and empowering management team, or some well-publicized adverse event that highlights the organizational penalties of silence.
- *Workarounds:*[7] Viewed from a managerial perspective, it may seem that a complex socio-technical system functions adequately because the workforce follows standard operating procedures. This is part of the story—but by no means all of it. When we examine the work of those on the front line, we can see that a significant part of it is concerned with solving local problems to get the job done. This daily massaging, tweaking, adjusting, compensating is largely invisible to top management. Nurses, in particular, obtain a good deal of professional satisfaction from overcoming the local difficulties of the hospital. If a piece of equipment is missing or malfunctioning, they may 'work around' the problem by taking a replacement from another location. This gets the job done, but

---

6 Morrison, E.W. and Milliken, F.J. (2000). Organizational silence: a barrier to change and development in a pluralistic world. *Academy of Management Review*, 25: 708–25.

7 Tucker, A.L. and Edmondson, A.C. (2003). Why hospitals don't learn from failures. *California Management Review*, 45: 55–72.

it means that those whose business it is to fix underlying organizational problems do not get to hear of these difficulties, and so an opportunity for system improvement is lost.

- *The normalization of deviance:*[8] This is a phrase coined by the American social scientist Dianne Vaughan in her account of the Challenger shuttle disaster in January 1986. It describes the process whereby certain defects become so commonplace and so apparently inconsequential that their risk significance is gradually downgraded, so that they are seen as routine wear and tear. The immediate cause of the Challenger accident was the explosive rupture of an O-ring on the launch rocket. Erosion of the O-ring had been noticed on several prior launches, but these incidents were not seen as signals of danger, largely because no bad event had occurred. What the NASA engineers had not appreciated was that the brittleness of the O-ring would be catastrophically enhanced by the unusually low temperatures prevailing at the time of the Challenger launch. The same process was invoked by the Columbia Accident Investigation Report as a main contributor to the Columbia shuttle tragedy in January 2003. Foam debris damage to the wings of the spacecraft had been observed on all 113 prior shuttle flights, but it was discounted as not constituting a serious risk. In the event, it was a large piece of foam that penetrated the wing structure and caused its destruction on re-entry. History had apparently repeated itself.

8 Vaughan, D. (1996). *The Challenger Launch Decision*. Chicago: University of Chicago Press.

Cultural Strata

It would be useful to conclude this chapter by describing Patrick Hudson's extension of Ron Westrum's typology of organizational safety cultures. Patrick takes an evolutionary view of these stages, believing that each has to be passed through before moving on to the next level. They are listed below:

- *Pathological*: Blame, denial and the blinkered pursuit of excellence (VSS). Financial targets prevail: cheaper/faster.
- *Reactive*: Safety given attention after an event. There is concern about adverse publicity. The organization establishes an incident reporting system.
- *Bureaucratic or calculative*: There are systems in place to manage safety, often in response to external pressures. Data harvested rather than analyzed and used. Safety management is strictly 'by the book'.
- *Proactive*: Aware that 'latent pathogens' and 'error traps' lurk within the system. Seek to eliminate them before they combine to cause an accident. Management listens to experts and 'sharp enders'.
- *Generative (high-reliability)*: Respects, anticipates and responds to risks. A just, learning, flexible, adaptive, prepared and informed culture. Strives for resilience.

In the next two chapters we focus on the health care domain. Patient safety has been my primary concern since the mid-1990s.

Chapter 12
Medical Error

Introduction

It is convenient to date the current widespread concern with patient safety from the beginning of the millennium. That was when the enormously influential US Institute of Medicine (IOM) report 'To Err is Human'[1] was published. Other patient safety issues had preceded it, particularly among anaesthetists, but it was the IOM report that galvanized the attention of a broader healthcare readership. This report was quickly followed by a series of high-level publications from a number of countries.[2] Unlike in other domains, there was no widely publicized 'big bang' event that engendered this concern, just these widely read national reports that detailed the huge costs of adverse events in health care, in terms of both lives lost and money expended. The numbers were staggeringly large: approximately one in ten (plus or minus two) patients in acute care hospitals were killed or injured as a consequence of medical error or institutional shortcomings. A recent French report puts this number as high as 14 per cent.

---

1 Institute of Medicine (2000). *To Err is Human: Building a Safer Health System*. Washington, DC: National Academy Press.

2 See, for example, Donaldson, L. (2000). *An Organisation with a Memory*. London: Department of Health.

The scope of the patient safety problem is huge. But there is nothing specifically 'medical' about medical errors. The only thing uniquely medical about them is the context in which they occur: doctors, nurses and pharmacists are—just like the rest of us—human and therefore fallible.

The Paradox

There is a paradox at the heart of the patient safety problem. Like all paradoxes it has at least two contradictory elements.

The Paradox: Part One

Health care training, particularly that of doctors, is predicated on a belief in trained perfectibility. After a very long, arduous and expensive education, you are expected to get it right. The consequence is that medical errors are marginalized and stigmatized. They are, by and large (in stark contrast to those in aviation and many other domains) equated to incompetence. As a result there has been no tradition of reporting and learning from errors. Doctors know they are fallible, but—until recently—they had no opportunity for sharing with other professionals. So, many responsible doctors resorted to noting down their errors in 'little black books', and their records remained personal. It is also the case that most medical students (in the past) did not learn about error-producing situations: error traps, situations in which the same circumstances produce the same or very similar errors in different people. A recent study of surgeons conducting a complex paediatric cardiovascular procedure on infants (the switch operation) indicated that the virtuosi were not those who never made errors (they all did), but those who detected and corrected their mistakes. They were the ones who had the best outcomes.[3]

The Paradox: Part Two

If an evil genius were given the task of contriving the most error-productive activities on the face of the planet, he (or perhaps she) would come up with two: aircraft maintenance and delivering health care. The following bullet points summarize the argument with regard to health care. In so doing it is useful to contrast health care with in-flight aviation (something that is offered as a contrast to medicine as a model of safety management). I do not dispute that it is a useful model, but only if one can appreciate the very large differences between the two domains.

- *A huge diversity of activities and equipment*: Long-distance pilots fly predominately two types of aircraft: Boeing and Airbus. Health care professionals, especially nurses, may encounter up to 40 different infusion pumps in one hospital, many with quite different calibrations. In addition, their activities cover an enormous range of therapeutic behaviours.
- *Hands-on work with limited safeguards*: Pilots, by and large, are not encouraged to touch the flying controls. In the 13 hours between Singapore and London, for example, they may be handling the

3 Carthey, J., de Leval, M. and Reason, J. (2001). Assessing the resilience of health care institutions to the risk of patient mishaps. *Quality in Health Care*, 10: 29–32.

controls for 15 minutes. But health care work is very largely hands-on. The more you touch, the greater the opportunity for making errors. Unlike other domains (aviation, nuclear power, chemical process, etc.) there are relatively few engineered defences. All that separates, for example, a surgeon's scalpel from cutting a wrong nerve or blood vessel is his or her skill.

- *Vulnerable and needy patients*: Unlike passengers, patients are sick or injured. Errors are far more likely to have damaging effects. But then there is the 'lethal convergence of benevolence'. Health professionals care about their patients. They do whatever is necessary to make their lives easier. That often means bypassing the protocols, barriers and safeguards.
- *Local event investigation*: With few exceptions, adverse events do not reach the public arena. Instead they are investigated locally. This means that the lessons learned are not widely disseminated.
- *One-to-one or few-to-one delivery*: Health care is up close and personal, unlike many other activities. As such, the potential for error is much greater.

The paradox in a nutshell: health care by its nature is highly error-provoking—yet health carers stigmatize fallibility and have had little or no training in error management or error detection.

Models of Medical Error

Error dominates the risks to hazardous domains, especially in health care. There are two main perceptions of error: the person model and the system model. Each has its own theory of error, and each directs a particular type of remedy. They have been discussed in detail elsewhere.[4] In addition, there are two particularly unhelpful models: the plague model and the legal model. We will deal with those first.

The Plague Model

One reaction to the discovery of a high incidence of errors in health care institutions has been to describe them as 'a national problem of epidemic proportions'. The term 'epidemic' puts error in the same league as AIDS, SARS or the Black Death (*Pasteurella pestis*). A logical step from this 'plague model' is to call for the elimination of human error; but, unlike some epidemics, there is no specific treatment for error. It is part of the human condition. While we can't fundamentally change this condition, we can change the conditions under which health care professionals work to make them less error provoking.

A closely related view is that errors are the product of built-in deficiencies in the human condition—a perspective that leads system designers to strive for ever more advanced levels of automation in order to keep fallible people out of the loop as far as possible.

The problem stems from confusing error with its occasional consequences. That errors can and do have adverse effects leads people to assume that human error is a bad thing. But this is not the case, as discussed in earlier chapters.

4 Reason, J. (1997). *Managing the Risks of Organizational Accidents*. Aldershot: Ashgate.

The Legal Model

This is a variant of the person model but with strong moral overtones. Central to this view is the belief that responsible and highly trained professionals should not make errors. They have a duty of care. Those errors that do occur are thought to be few but sufficient to cause adverse effects. Errors with bad consequences are assumed to be negligent or even reckless and thus deserve deterrent sanctions—echoes of the 'just world' hypothesis.

Far from being rare occurrences, the research evidence shows that highly trained professionals make frequent errors, but most are inconsequential or detected and corrected. A study involving direct observations of 165 arterial switch operations (surgically correcting the congenital transposition of the aorta and the pulmonary artery in neonates) in 16 UK centres showed that, on average, there were seven errors in each procedure, one of which was major and life-threatening, while the others were little more than minor irritations. Over half of the major errors were detected and corrected by the surgical team.

Neonatal cardiothoracic surgery takes surgeons to the very limits of their skills: it is lengthy and highly complex. Errors are inevitable. The virtuosi surgeons make errors, but they anticipate them and prepare themselves to detect and recover them. As we have seen, this mental and organizational preparedness—or mindfulness of danger—is one of the defining characteristics of high-reliability organizations.

Clearly, error-making is the norm, even among highly accomplished professionals. Training and experience do not eradicate fallibility; they merely change the nature of the errors that are made and increase the likelihood of effective compensation. The record shows that, far from being sufficient to cause bad outcomes, only very occasionally are the unwitting acts of professionals necessary to add the final touches to a disaster that has been waiting on the sidelines, often for a lengthy period. What follows is a brief account of the two most dominant models of error: the person model and the system model.

The Person Model

This is the most widely held view of errors, particularly in health care. It places their origins squarely between the ears of people at the sharp end, the patient–professional interface. It sees errors as the product of wayward mental processes: forgetfulness, inattention, preoccupation, distraction, ignorance, carelessness and the like.

Remedial measures derived from this model are directed primarily at the sharp-end error maker in an effort to change the person's motivation and to prevent the recurrence of future errors. The methods include naming, blaming, shaming, retraining, fear appeals and writing another procedure. Although intuitively appealing—blaming is such a powerful and satisfying emotion—it is largely ineffective; indeed, it can be counterproductive. Managers like it because it separates the errant individual from the organization in which he or she works. Lawyers like it because it fits neatly into the tort law structure. But it also isolates the person from the context in which the error was made and impedes the discovery of error traps—similar situations produce similar errors in different people. That is, we are talking about error-prone situations rather than error-prone people.[5]

It was largely to combat this outmoded but persistent view of error making that the Institute of Medicine (IOM) report was published. The IOM report strongly endorsed the system view of error, to be discussed below.

The System Model

The basic premise here is that humans are fallible and that errors are to be expected, even in the best organizations. They are mostly not a moral issue. Errors are seen as consequences rather than causes, having their origins not so much in the perversity of human nature as in 'upstream' systemic factors. These include recurrent error traps in the workplace and the organizational processes that give rise to them. A central idea is that of system defences. All hazardous technologies possess barriers, defences and safeguards to guard against known dangers. When an adverse event occurs, the important issue is not who blundered, but how and why did the defences fail.

The patient safety movement, shaped mostly by the system model, has been going strongly for a dozen years. Has it made a real difference?

Conclusions

I have been a close lay observer of the patient safety movement since the late 1990s, and I have noted some significant steps forward. Foremost, I think, is the way that the patient safety issue has greatly

5 Reason, J. (1990). *Human Error*. New York: Cambridge University Press.

increased in salience and significance among health care professionals and, to some extent, their patients. Since 2000, for example, I have attended some 110 conferences and meetings relating to patient safety in several countries. There have also been many excellent books, journals and journal issues devoted exclusively to the topic.

At the moment, we cannot mark the success of the movement by a dramatic decrease in harmful medical errors across the board—but there have been pockets of real progress in anaesthesiology, surgery, ward design, use of checklists, clean hands policies and in many other areas. Carers have come increasingly to recognize the importance of communication, briefing and team-working skills (largely due to the intervention of a group of dedicated commercial pilots and other human factors experts). The patient safety problem has been recognized as one of the most challenging, important and intellectually interesting research topics in town. I once heard a speaker say, 'Safety management, that's not rocket science.' He was right. Rocket science is trivial compared to managing safety—in any domain. I have been privileged to meet many, if not most, of the leading patient safety players in the United States, Canada, Australia, New Zealand, Finland, Denmark, Sweden, Italy, Ireland and the UK. We have become a tightly knit international community with a clear identity and a shared purpose.

But professionals remain human and thus fallible, and the business of delivering health care continues to be highly error provoking. If we haven't cracked the problem, what have we done? Most of our successes relate to the way we regard medical errors rather than their actual reduction. Dr Lucian Leape, the 'godfather' of patient safety, once said that you will know that the patient safety movement has made progress when you enter a ward, a clinic or an operating theatre and hear professionals talking spontaneously about the problem and the ways to manage it. I think that could certainly be true in big city teaching hospitals. I am less sure about the smaller town institutions and primary care practices.

I also think that the nature of medical error is much better understood now than it was 20 years ago. We appreciate that errors do not occur in isolation but have traceable systemic origins. We know that we cannot eliminate error entirely. It is deeply rooted—for good or ill (and not always the latter[6])—in the human condition. We can't change the human condition, but we can change the conditions under which health carers work. That means changing the culture of medical education and health care delivery. It means removing known 'error traps' and encouraging a climate of learning. Each investigated medical error is an opportunity to improve our understanding.

Maybe we have got as far as we could reasonably get in just over a decade, given the complexity of the domain. It is quite understandable for some health care professionals to think that everything they do is about safety. But it isn't. Managing safety like any other aspect of health care is a skill that requires training and practice.

So, progress in patient safety is a maybe, but a largely optimistic maybe. This is something that will take many decades to achieve—and there will be no final victory, just, one hopes, some significant and sustained improvement. Managing safety is like fighting a guerrilla war—it's always one damn thing after another.

---

6 Ibid.

Chapter 13
Disclosing Error

This penultimate chapter discusses an important issue relating to medical error: Should healthcare professionals tell their patients that they have made an error—even when it has only minor consequences? In short, should they say 'sorry'? Or would this increase the number of potential litigants?

In February of 2011, I presented a two-part BBC Radio 4 series entitled *Doctor, Tell Me the Truth*. Many British and American experts contributed, along with the relatives of a number of harmed patients. This brought the error journey up to date.

For the most part, I will focus on developments in the United States. Why? Because the US is arguably the largest, the most advanced, the most complex and perhaps still the most litigious health care system in the world. It is also one of the most transparent in the sense that most of the crucial issues are discussed in the public domain. It has also led research in this area.

Reporting Systems

The IOM report not only re-energized the nascent patient safety movement in the United States and elsewhere, it also set in motion a train of events that led directly to the disclosure issue. Among its eight chapters (and one appendix), there were two devoted to error reporting systems and the need to protect them from legal discovery. The publication of the IOM report coincided with a spike in malpractice premiums and an increase in the number of state mandatory reporting systems. This was also associated with an increased demand from doctors in these states for stronger protection of reported data.

Reporting systems whose main purpose is to hold health care providers accountable are termed 'mandatory reporting systems'. They focus upon errors associated with serious injuries or death. Most mandatory reporting systems are operated by state regulators who have the power to investigate specific cases and, where necessary, issue sanctions or fines for wrongdoing. These systems serve three purposes.

- They provide patients with a minimum level of protection by assuring that serious errors are reported and investigated and that appropriate follow-up action is taken.
- They provide an incentive to health care organizations to take a systemic approach to patient safety so as to avoid penalties and public exposure. The system approach (in stark contrast to the tort system) seeks to bring mistakes out into the open in order to fix organizational flaws and thus avoid future lapses. The tort system, on the other hand, blames an individual for his or her wrongdoing, and punishment seeks to deter future errors. It is now widely agreed that the tort system does not achieve its purpose, either in adequately recompensing patients who have suffered adverse events or in preventing their recurrence.

- Mandatory reporting systems require all health care institutions to make some level of investment in patient safety. While errors resulting in serious harm are but the 'tip of the iceberg', they signal major system breakdowns with serious consequences for patients.

The IOM Committee strongly believed that mandatory reporting systems should be coupled with voluntary reporting systems that focus on errors having minimal or no harm. Analyses of these 'free lessons' provide valuable information for those who seek to improve the resilience of health care systems in general by making errors less likely, more readily detectable and less harmful. Without such systems, the organization has no 'memory' and hence only a very limited ability to learn from its mistakes.[1]

Obstacles to Reporting

Health care workers, and especially doctors, have traditionally had little sympathy for or understanding of medical errors. They are marginalized, stigmatized and equated with technical incompetence. It is not surprising that the state and IOM pressures to report such events filled many surgeons and physicians with the fear that increased reporting would lead to an increase in malpractice suits. It runs counter to the deeply ingrained tradition (shared by many doctors, hospital managers and legal advisors) of 'deny and defend'. But so far there is no evidence of a causal

---

1 Donaldson, L. (2000). *An Organisation with a Memory*. London: Department of Health.

relationship between mandatory occurrence reporting and increased malpractice litigation.[2]

The temptation not to report a medical error is very strong. One study[3] revealed that only 50 per cent of those house physicians who admitted making serious clinical errors disclosed their errors to medical colleagues, and only 25 per cent disclosed them to the affected patients or their families. Another study involving patients reported that only a third of the physicians involved in the error had informed them about it.[4] A European survey[5] found that although 70 per cent of physicians believed that they should disclose details of their errors, only 32 per cent actually did. Other researchers have emphasized that being subjected to a malpractice lawsuit is 'an extremely powerful punishment that strikes at the heart of the professional's self-image as a caring and competent physician'.[6]

Apologies and Compensation

Mere disclosure of medical error is one thing; apology is quite another. Whereas disclosing has been

---

2 Marchev, M. (2003). *Medical Malpractice and Medical Disclosure: Balancing Facts and Fears*. Portland, ME: National Academy for State Health Policy.

3 Wu, A.W. et al. (1991). Do House Officers Learn from Their Mistakes? *JAMA*, 265: 2089–94.

4 Blendon, R.J. et al. (2002). View of practicing physicians and the public on medical errors. *New England Journal of Medicine*, 347: 1933–40.

5 Vincent, C. et al. (1994). Why do people sue doctors? A study of patients and relatives taking legal action. *Lancet*, 343: 1609–13.

6 Runciman, W.B. et al. (2003). Error, blame and the law in healthcare—an antipodean perspective. *Annals of Internal Medicine*, 138: 974–9.

advocated for a long time, the practice of offering an apology is relatively new. The question of whether errors committed by doctors should be disclosed to patients affected by them is no longer debatable. The weight of legal opinion, state regulators, accreditation agencies and the physician's professional code of ethics all favour the complete disclosure of the factors relevant to the patient's health. But the question of what doctors should tell patients—in other words, how they should apologize—remains a complicated issue, as will be discussed later.

The Veterans Affairs Medical Center (VAMC) in Lexington, Kentucky, normally gets the credit for setting the apology ball rolling. In 1987 the VAMC management adopted an untried and untested approach to medical errors. It has been called the 'disclose-apologize-compensate' algorithm. It stands in stark contrast to the age-old 'deny and defend' strategy. The management decided that it needed a more proactive approach to medical error after losing two malpractice claims totalling $1.5 million.[7]

In 1999, 12 years later, Dr Kraman and his colleagues[8] reported that the practice of disclosing, apologizing and compensating for medical errors continued to be followed because the staff believed it was the right thing to do. It also resulted in unanticipated financial benefits to the medical center. The team reported that its disclosure practice suggests but does not prove the financial superiority of a full disclosure policy. In 1999, the Lexington facility also compared its claims history with those of 36 similar VAMCs located east of

7 Marchev, M. (2003).

8 Kraman, G. and Hamm, G. (2007) Letter to the Editor. *Health Affairs*, 26, 3 June.

the Mississippi River. They modestly concluded that Lexington VAMC's liability payouts were moderate and comparable to those of similar centers. It ranked 29th out of 36 for dollars paid.

All 36 centers provided tertiary care and were closely linked to medical schools. The study estimated risk exposure by comparing 'the complexity-adjusted workload' for each of the 36 facilities. This revealed that VAMC Lexington ranked 23rd out of 26 on this measure. In other words, Lexington ranked 23rd for risk exposure, but ranked 29th in dollars paid.

In a 2001 publication Dr Kraman stated, 'We average about 14 cases a year. That's high for a VA hospital. But the average payment per case is only $15,000. The average per case for the VA system as a whole is closer to $100,000'. In a 1999 paper, they concluded, 'An honest and forthright risk management policy that puts the patient's interests first may be relatively inexpensive because it avoids the costs of lawsuit preparation, litigation, court judgements and settlements at trial. Although goodwill and maintenance of the caregiver role are less tangible benefits, they are also important benefits of such a policy.'

In her excellent paper, from which many of the details of the Lexington study have been taken, Nancy Lamo[9] suggests that both the IOM and Lexington reports have brought about a cultural change as indicated by the arrival in the medical lexicon of terms like transparency, disclosure and apology. In 2001, the Joint Commission made disclosure a requirement for accreditation. Interestingly, though, they stopped short of requiring an apology in the belief that it

---

9 Lamo, N. (2011). *Disclosure of medical errors: The right thing to do, but what is the cost?* Lockton Companies.

would increase malpractice claims. In the same year, the University of Michigan Health System (UMHS) began a very sophisticated program of research on disclosure—to be discussed shortly.

A high point in this cultural evolution occurred in September 2005, when Senators Hillary Clinton and Barack Obama proposed a bill to establish the National Medical Error Disclosure and Compensation program that built on the Patient Safety and Quality Improvement Act which passed into law in July 2005. The program seeks to reduce malpractice lawsuits by providing liability protection for health care providers who rapidly disclose medical errors, offer apologies and are prepared to enter into negotiations for fair compensation. As of 2006, the bill is still pending review by the Senate Committee on Health, Education, and Pensions.

By the end of 2005, apology-immunity statutes had been passed in 19 US states. Illinois had a 'Sorry works' law that was taken up by the UK's National Patient Safety Agency (NPSA), which translated it into the 'Being Open' Policy, which was intended to be fully implemented by the end of 2006. But since then the NPSA has been disbanded by the Coalition government; who knows what has become of it?

Not all US states were equally generous with their immunity policies. Some do not protect statements that admit wrongdoing of one kind or another. Colorado, for example, offers immunity for apology but not for admissions of fault. Such differences place a great emphasis on how an apology is framed and delivered. As Professor Albert Wu of Johns Hopkins University has pointed out, it's not so much what is said but what the recipient hears that matters.

Lee Taft, an ethicist,[10] has argued that for an apology to be authentic it must contain the following elements:

- An acknowledgement that a rule or protocol has been violated.
- An expression of genuine remorse and regret for any harm caused.
- An explicit offer of restitution.
- A promise of reform.

The manner of disclosure has become an important research issue. Many argue that disclosure should not be left to a single individual, no matter how well-intentioned or articulate he or she is; rather, it should be a formalized team effort. Another model is disclosing through the medium of peer-reviewed journals (for example, the 'Uses of Error' column in the *Lancet*); yet another is professional society error reporting. And then, of course, there is the University of Michigan model described below.

The University of Michigan Health System (UMHS) Program: A Before and After Study

The purpose of the study was to compare liability claims and costs before and after the implementation of the UMHS disclosure-with-offer program. Since 2001 UHMS has fully disclosed and offered compensation to patients with medical errors. They linked two data sets. One was the risk management database containing claims-related performance data,

10 Berlin, L. (2006). Will saying 'I'm sorry' prevent a malpractice lawsuit? *American Journal of Roentgenology*, 187: 10–15.

such as injury and disposition dates, disposition status and liability costs. The second was the clinical information and decision support service database. The four primary study measures were the number of new claims, number of claims receiving compensation, time to claim resolution and claims-related costs. The study reviewed 1,131 claims reported to UHMS risk management from July 1995 to September 2007. The results are summarized below:[11]

- *Fewer lawsuits*: There were 38.7 lawsuits per year before and 17.0 per year after disclosure program implementation. The monthly rate of new claims decreased from 7.03 per 100,000 patient encounters to 4.52 after full implementation.
- *Faster resolution*: The median time to resolution was 1.36 years before initial implementation, and 0.95 year after initial program implementation.
- *Lower costs*: The average cost per lawsuit decreased from $405,291 before to $228,308 after program implementation. The mean legal expenses for UHMS decreased by about 61 per cent, but part of these savings were offset by the increase in the UMHS risk management budget needed to more proactively address claims internally.

Mr Richard Boothman,[12] a lawyer and risk manager as well as a prime mover in the UHMS project, argued strongly that the reason for disclosure of medical errors is not so much cost savings, but the larger goal

11 Lamo, N.

12 Boothman, R. (2010). Message on National Patient Safety Foundation Listserv (20 November).

of patient safety. Honesty is the basis of the three principles embodied in the UHMS study.

First, when inappropriate medical care causes the patient harm, the provider owes the patient quick and fair compensation. Second, where there was no patient injury, the caregivers are due a thoughtful and vigorous defence. Third, underlying it all is the need to learn from patients' experiences to improve the quality of care. Although it is hard to contest the rightness of these three principles relating to honesty and disclosure, there still remain some powerful naysayers.

Contrary Voices

In 2007, Dr David Studdert was the lead author on a paper entitled 'Disclosure of Medical Injury to Patients: An Improbable Risk Management Strategy'.[13] The paper is significant not only for its contrary view, but also because it includes at least two major figures in patient safety among its authors. One is the surgeon and essayist Dr Atul Gawande, a prime mover in the development and dissemination of the surgical checklist. The other is Dr T. Brennan, the first author of the Harvard Medical Practice Study (1991) that estimated that between 44,000 and 98,000 Americans are killed or injured each year as the result of medical error. This was one of the key studies that triggered the IOM report nine years later. These are voices that cannot be easily dismissed.

Using data from 65 medico-legal experts, the study concluded that '... the chance that disclosure would decrease either the frequency or cost of malpractice

13 Studdert, D. et al. (2007). *Health Affairs*, 26: 215–26.

litigation is remote. On the contrary, an increase in litigation volume and costs was highly likely.' These conclusions were based on the following observations:

- The vast majority of patients suffering as the result of medical error never sue. This results in a huge number of potential claims. Only 2 to 3 per cent of patients injured by negligence actually file malpractice claims.[14]
- One reason why such patients do not sue is that they are unaware they have been victims of medical error.
- As a result, the authors argue, disclosure of medical error will prompt claims. Even a small increase in the number of claims prompted by disclosure will swamp the relatively small number of claims that may be deterred by honest disclosure.

This study gave rise to a storm of protest. The VAMC authors described it as a flawed study that was both 'irresponsible and bad science'.[15] Similar criticisms arose across the world, particularly in Australia. The Australians noted that it is simply not proven that patients' unawareness of the cause of their injuries is a significant factor in explaining why they do not sue. It is known that patients instigate litigation because they suspect a cover-up.[16]

---

14 Mello, M. et al. (2007). An analysis of adverse events costs, the medical liability system and incentives for patient safety improvement. *Journal of Empirical Legal Studies*, 4: 835–60.

15 Kraman and Hamm.

16 Lamo, N.

Conclusions

So where do we currently stand on the disclosure issue? In my view, the bulk of the evidence favours the 'disclose-apologize-compensate' model introduced by the VAMC Lexington and UMHS studies. Both covered lengthy time periods and both were rigorously conducted. But the naysayers also have a strong point. As yet, however, we do not have enough data to resolve these competing claims. One thing is certain: the traditional 'deny and defend' strategy is unacceptable in the twenty-first century.

Chapter 14

Reviewing the Journey

This journey began in the early 1970s with a bizarre slip in my Leicestershire kitchen and has concluded in health care institutions and the law courts. En route I have worked in a wide variety of hazardous domains—aviation, nuclear power installations, chemical process plants, railways, the maritime world, oil and gas production, banks, road transport, mining, dams, air traffic control and a few others.

The journey has also taken me to many continents and several countries. Along the road, I have discovered two things: learn as much as possible about the details of how people in these various domains work, and, most important, never be judgemental. These are the messages I have tried to persuade my students to follow. Which brings me to my next point: since 1964, I have been employed full-time by two universities: first, the University of Leicester, and for the past 36 years, my *alma mater*, the great University of Manchester. I have no complaints. Not a bad life for an academic.

I have to make an apology. It is inevitable that in reviewing a lifetime's research I will occasionally repeat what I have written somewhere else. I have tried to avoid plagiarizing myself, but I have not always succeeded. For this I am sorry. Though I think it helps to think of a book as a self-contained unit. So, reminders are often necessary; nevertheless, my apologies.

The years following the cat food incident were occupied with unravelling the components of error using self-report questionnaires and diary studies. My mentors at this stage were Jens Rasmussen in Denmark, Don Norman in California, Berndt Brehmer in Uppsala and Donald Broadbent in Oxford. In regard to the latter, members of the Department of Psychology in Oxford were a notoriously tough audience of experimentalists. I expected to be booed off the stage when I gave a talk there. But I wasn't—I think this had to do with the fact that everyone recognizes absent-minded actions as being part of their own repertoire of actions. They are self-validating to a certain extent.

This period, the late 1970s and 1980s, was taken up with classification and with making various distinctions between the varieties of unsafe acts. These early categories are listed below.

- Slips and lapses versus mistakes—the first distinction.
- Rule-based and knowledge-based mistakes—the second distinction.
- Errors and violations—the third distinction.
- Active versus latent human failures—the fourth distinction (though the term 'latent failures' was later changed to 'latent conditions').

It was at this point that I ceased being a kosher cognitive psychologist and started working in the real world of hazardous industries. I was dragged out of the ivory tower by my friend David Embrey, an applied psychologist with an engineering background.

There was a lot going on in the real world. The distinction between active and latent failures owes a great deal to Mr Justice Sheen's observations regarding the capsize of the *Herald of Free Enterprise* in 1987. He wrote:

> At first sight the faults which led to this disaster were the ... errors of commission on the part of the Master, the Chief Officer and the assistant bosun ... But a full investigation into the circumstances of the disaster led inexorably to the conclusion that the underlying or cardinal faults lay higher up in the Company ... From top to bottom the body corporate was infected with disease of sloppiness.[1]

In short, the *Herald* was a 'sick' ship even before it sailed from Zeebrugge on 6 March 1987. The latent failures included:

- Low-capacity ballast pumps that were insufficient to put the vessel on an even keel in less than half the voyage time. Requests for higher-capacity pumps had been rejected.
- Despite repeated warnings, there were no remote bridge indicators to detect and warn of open bow doors.
- The vessel had a chronic list to port. This, together with the inadequacy of the scuppers and the high centre of gravity, made it inevitable that the ship would turn turtle rapidly and catastrophically.
- Inadequate checking of the number of passengers aboard. Numbers had exceeded passenger limits on several prior occasions.

1 Sheen, Mr Justice. (1987). MV Herald of Free Enterprise. Report of Court No. 8074. Formal Investigation. London: Department of Transport.

- Poor storage and design of lifejackets. Tangled masses of floating lifejackets prevented some swimmers from reaching the surface.
- High crew workload due to undermanning and poor tasking.
- Low morale due to poor working conditions and longstanding disputes between seafarers and land-based management.
- Company procedures that were both ambiguous and inappropriate.
- Safety issues were subordinated to those of productivity and cost-cutting in a wide range of activities.
- A culture that condoned (or rendered unavoidable) violations of safe operating procedures.

The distinction between active and latent failures rests upon two considerations: first, the length of time before the failures have a bad outcome; second, where in the organization the failures occur. Generally, active failures are committed by those in direct contact with the system and have an immediate though often short-lived impact; latent failures occur within the higher echelons of the company, and their adverse effects may be delayed by many years.

The latent failure idea got a huge boost from an unexpected direction. In 1986–87, I was writing a book called *Human Error* for Cambridge University Press. When the first draft was complete, I asked my friend Berndt Brehmer, then professor of psychology at the University of Uppsala, to read it critically. This he did and came back with the annotated draft. He told me he liked most of it well enough, but Chapter 7 was boring and should be omitted. It traced the history of error from the ancient Greeks onward. It had taken a long time to prepare and I quite liked it. But I could see his point. Dumping it, however, left a big gap toward the end of the book. We were at that time in the middle of a cluster of well-investigated organizational accidents (see Chapter 10). So it seemed appropriate to replace it with a chapter entitled 'Latent Errors and System Disasters'. And that was how the Swiss cheese model was born—I didn't call it that, though I am eternally grateful to Dr Rob Lee, then Director of the Bureau of Air Safety Investigation (BASI) in Canberra, who actually came up with the winning label. All I did was produce a set of graphics that had slices (defences) with holes in them. Without the catchy title and the new chapter, the book would probably have been judged as overly academic—it was written for psychologists—and too full of psychobabble. Interestingly, the sales of this book defied gravity and increased around 2000, and it is still selling well. My belief was that health care professionals were buying it—which I now know to be the case. Swiss cheese had a new lease on life after the patient safety movement took off in the early years of the millennium. It was recently described as 'the dominant paradigm for analyzing medical errors and patient safety incidents'.[2] Googling 'Reason Swiss cheese' produced 2,560,000 hits. It's a very crude index, but it does suggest that Swiss cheese got out and about a bit.

One thing is clear: a journey like this does not go very far without help from a large number of people. I never had a 'group' as such in Manchester

---

2 Perneger, T. (2005). The Swiss cheese model of safety incidents: Are there holes in this metaphor? *BMC Health Services Research*, 5: 71.

(although research assistants, research students and many undergraduates came and went over the years). My 'close others' were scattered across the globe: Don Norman and Najm Meshkati in California; John Wreathall and Dave Woods in Ohio; Richard Cooke in Chicago; Jens Rasmussen in Denmark; Dietrich Doerner in Germany; Carlo Cacciabue in Italy; Jan Davies in Canada; Rob Lee and Andrew Hopkins in Australia; Lucian Leape and Don Berwick in Massachusetts—to name but a few. Closer to home, I got invaluable guidance from Marc de Leval, Charles Vincent, Dianne Parker, Rebecca Lawton, Jane Carthey and Carl Macrae. Another friend and longstanding ally was the recently retired Chief Medical Officer, Sir Liam Donaldson. But my staunchest helper and advisor was my wife (and management), Rea, also a psychologist.

This will be my valedictory error book: I have nothing new to say and I'm well past my best. No more cat food in the teapot—we have outlived our cats—but I still remain the most absent-minded person I know, so I'm sure I will find some new nonsense to commit. Thank you for your interest and support. Be safe.

Postscript

A few days after I had sent the bulk of this book to the publisher, Mr Robert Francis published the 'Report of the Mid Staffordshire NHS Foundation Trust Public Inquiry'.[1] This troubled hospital had been investigated several times before and its parlous condition had been known about for nearly a decade, but Mr Francis led the first statutory inquiry (pursuant on the Inquiries Act 2005) that is available in the public domain. Since so many of the issues discussed in the report closely echoed themes to be found in the latter part of this book, I thought it worthwhile to comment briefly on some of them here.

Mr Francis counted up the number of times the word 'hindsight' cropped up in transcripts of the oral hearings of the Bristol Royal Infirmary. It occurred 123 times and 'the benefit of hindsight' 378 times. If, for quite different reasons, a count were to be made of the occurrence of the word 'culture' in this inquiry report, I am sure it would greatly outnumber the figures above. Indeed, nearly the whole of the inquiry report, in one way or another, is about the negative aspects of culture in the Mid Staffordshire system. These are summarized below:

- A lack of openness to criticism
- A lack of consideration for patients

1 London: The Stationery Office, February 2013.

- Defensiveness
- Looking inwards, not outwards
- Secrecy
- Misplaced assumptions about the judgements and actions of others
- An acceptance of poor standards
- A failure to put the patient first

Factors shaping these characteristics are discussed in Chapter 11. The inquiry report makes this comment: 'During the course of both the first inquiry and the present there has been a constant refrain from those charged with managing, leading, overseeing or regulating the Trust's provision of services that no cause for concern was drawn to their attention or that no one spoke up about concern.' The issues that block bad news from travelling upwards are discussed at length in Chapter 11.

This chapter also describes the 'vulnerable system syndrome' made up of three mutually potentiating pathologies: blame, denial and the blinkered pursuit of the wrong kind of excellence. All three are evident in this report, but the latter comes into particular prominence: 'Management thinking during the period under review was dominated by financial pressures and achieving foundation trust status to the detriment of quality of care.'

My final comment concerns the close connections between Mr Francis's remarks at various points in the report and the penultimate chapter here on disclosing harm to patients. In his words: 'Ensure openness, transparency and candour throughout the system about matters of concern.' I have heard the media using the phrase 'duty of candour' with the implication that it should have the same legal status as 'duty of care'.

Index

Printed in the United States
by Baker & Taylor Publisher Services